Third Edition

Clinical Case Discussion in Biochemistry

Based on NExT-I Pattern of Examination

A Book on Early Clinical Exposure (ECE)

As per the latest CBME Guidelines | Competency Based Undergraduate Curriculum for the Indian Medical Graduate

Book Review for *Clinical Case Discussion in Biochemistry*

Dr Poonam Agrawal

"My gratitude to this book and the amazing author of this masterpiece–our HOD, Department of Biochemistry–Dr Poonam Agrawal Ma'am is something that I would never forget and cherish for my whole life as a future doctor.

This book discusses hundreds of clinical cases of medical world with important biochemical basis for their clinical signs and symptoms, that helped me to secure distinction in subject of biochemistry for first professional university examination and make my concepts crystal clear for future as well, enlightening me on the path of becoming a good doctor.

I truly appreciate the immense hard work and time that ma'am spent on writing this masterpiece and would encourage other students too, to read this book and fall in love with the wonderful subject of biochemistry, once again.

Thank you so much ma'am, for all that you have done to make our medical studies fascinating and guiding us on the path for a bright future..."

Vyomesh Rastogi
3rd Year MBBS student
Dr BSAMCH, New Delhi
College Batch Topper and University Rank 3

The Clinical Case book by Dr Poonam Agrawal Ma'am is very good book. This book contains all the diseases that should be covered in the first year, up to this time we had read many diseases. The content of the book is easy to remember and can be written easily in exam which helps us to score good marks.

Ajay Singh
MBBS
Batch 2022

Clinical Case Discussion

I find Clinical Case book is one of the best books that I have studied, it helps to understand clinical cases. It gives me feel that we are slowly becoming doctors. These cases fascinates me a lot...

Alok Pandey
MBBS, 2022 Batch

Feedback of Clinical Case Discussion book

The main highlight of this book is that it covers variety of cases which we will be dealing with in our clinics. The diagrams and graphs in the book really helps with clear understanding of the disease and its diagnosis. The question–answer format of the book for every case makes it very interesting to read and makes understanding of every case clear.

Hardik Dhochak
MBBS
Batch 2022

Third Edition

Clinical Case Discussion in

Biochemistry

Based on NExT-I Pattern of Examination

A Book on Early Clinical Exposure (ECE)

As per the latest CBME Guidelines | Competency Based Undergraduate Curriculum for the Indian Medical Graduate

Poonam Agrawal
MBBS MD FIMSA ACME

Professor and Head, Department of Biochemistry
Head, Medical Education Unit
Dr Baba Saheb Ambedkar Medical College and Hospital
New Delhi

Additional Resource Faculty, FDP
Maulana Azad Medical College
New Delhi

CBSPD

CBS

CBS Publishers & Distributors Pvt Ltd

New Delhi • Bengaluru • Chennai • Kochi • Kolkata • Lucknow • Mumbai
Hyderabad • Jharkhand • Nagpur • Patna • Pune • Uttarakhand

Third Edition

Clinical Case Discussion in

Biochemistry

Based on NExT-I Pattern of Examination

A Book on Early Clinical Exposure (ECE)

ISBN: 978-93-5466-739-8

Third Edition: 2024

Reprint: 2025

First Edition: 2021

Second Edition: 2022

Reprint: 2023

Published by Satish Kumar Jain and produced by Varun Jain for

CBS Publishers & Distributors Pvt Ltd

4819/XI Prahlad Street, 24 Ansari Road, Daryaganj, New Delhi 110 002, India

Ph: 011-23289259, 23266861 Website: www.cbspd.com

e-mail: delhi@cbspd.com

Corporate Office: 204 FIE, Industrial Area, Patparganj, Delhi 110 092, India

Ph: 011-4934 4934 Fax: 011-4934 4935 e-mail: publishing@cbspd.com; publicity@cbspd.com

Branches

- **Bengaluru:** Seema House 2975, 17th Cross, K.R. Road, Banasankari 2nd Stage, Bengaluru 560 070, Karnataka, India
  Ph: +91-80-26771678/79 Fax: +91-80-26771680 e-mail: bangalore@cbspd.com
- **Chennai:** 18/8B, Subbarayan Street, Shenoy Nagar, Chennai 600 030, Tamil Nadu, India
  Ph: +91-44-42032115, 26681266 e-mail: chennai@cbspd.com
- **Kochi:** 42/1325, 1326, Power House Road, Opp KSEB, Power House, Ernakulam, Kochi, 682 018, Kerala, India
  Ph: +91-484-4059061-65 Fax: +91-484-4059065 e-mail: kochi@cbspd.com
- **Kolkata:** 147, Hind Ceramics Compound, 1st Floor, Nilgunj Road, Belghoria, Kolkata 700 056, West Bengal, India
  Ph: +91-33-25633055/56 e-mail: kolkata@cbspd.com
- **Lucknow:** Basement, Khushnuma Complex, 7-Meerabai Marg (behind Jawahar Bhawan), Lucknow 226 001, UP, India
  Ph: +91-522-4000032 e-mail: tiwari.lucknow@cbspd.com
- **Mumbai:** PWD Shed, Gala no. 25/26, Ramchandra Bhatt Marg, Next to JJ Hospital Gate no. 2, Opp. Union Bank of India, Noorbaug, Mumbai 400 009, Maharashtra, India
  Ph: +91-22-66661880/89 e-mail: mumbai@cbspd.com

Representatives

• **Hyderabad**	0-9885175004	• **Jharkhand**	0-9811541605	• **Nagpur**	0-8692091830
• **Patna**	0-9334159340	• **Pune**	0-9664372571	• **Uttarakhand**	0-9716462459

Printed at: Goyal Offset Works (P) Limited, Haryana, India *to*

---

my father
Dr Shri Prakash Agrawal and
my father-in-law
Mr Ram Baboo Agarwal

Preface to the Third Edition

I feel privileged to write Preface to the third edition of this book.

Clinical Case Discussion in Biochemistry was launched in 2021 as a novel and unique approach to fill the gap in biochemistry of clinical correlation of the subject and provide ECE (early clinical exposure) to healthcare students. Since that time this book has gone into two editions and reprints. In this edition, being launched in 2024, I am presenting completely revised and updated version of the book.

Many additional cases like ***cystic fibrosis, orotic aciduria, fatty liver in alcoholics*** are added on popular demand of faculties and learners from all over India.

In addition to new cases, the existing cases are strengthened with **additional questions** and their to-the-point explanations. I have taken this opportunity to add certain ***more illustrations like CT film of bronchiectasis, LDL receptors, mitochondrial DNA and its various elements, link of urea cycle and pyrimidine nucleotide biosynthesis in OTC deficiency, etc.*** to make concept more easy to grasp and for better retention.

I strongly believe that students find this edition much more useful than its previous editions and will help them to correlate biochemistry clinically in hospital and community set ups.

Feedback and suggestions are most welcome.

Poonam Agrawal
drpoonam24agrawal@yahoo.com
biochem25poonam@gmail.com

Preface to the First Edition

Clinical approach to biochemistry has been greatly emphasized for teaching this subject in medical colleges in **latest MCI guidelines on Competencies Based Medical Education (CBME) as per GMER regulation 2019.** Even in the assessment, the clinical understanding of the subject in form of **Case Based Questions (CBQs)** is greatly emphasized in latest curriculum.

In this context I felt the need of an exclusive book for medical undergraduates which **simplify the clinical understanding of biochemistry** and make the subject more relevant to learn in medical colleges.

The whole journey of writing this book has been an enriching experience. Case presentations are meticulously planned in such a manner that they are typical and close to actual clinical scenarios. CBQs are designed in such a manner as to generate the curiosity in learner mind and at the same time help in clear understanding the biochemistry basics.

While describing the clinical connections, special precaution has been taken to keep the **discussion simple,** keeping in mind that first year medical undergraduate student who is reading this book has no exposure to patients or hospital settings in the real world.

This book is first of its kind and I am confident that this book will be very easy to follow by the students. I strongly believe that learners find this book useful in making their clinical concepts clear and make them confident **to face Case Based Questions (CBQs) during their assessment.**

Feedback and comments are most welcomed and will be duly acknowledged in the next edition.

Wish you all the very best!!

Poonam Agrawal

drpoonam24agrawal@yahoo.com

You Tube link to my channel: *https://www.youtube.com/channel/UCzaWN-XIwROWrV-Aa8NBLiQ/*

Acknowledgments

First of all, I thank the Almighty for giving me strength and courage to work hard to best of my abilities and contribute to medical fraternity.

The whole journey of writing this book right from drafting the clinical cases along with the questions which is capable of giving greater insight to the biochemical aspect of the topic has been challenging but at the same time a thrilling experience.

I owe my sincere thanks and gratitude to many but first of all I wish to thank my husband **Dr Mohit Agarwal**, who has been a constant guide and motivator throughout the preparation of this book. His thoughtful insight has been instrumental in designing the book.

I appreciate my **daughter Misti** for her understanding and full cooperation when I was very busy in writing this book and could not completely assist her during her online classes. My pet **(peanut)** deserves a special mention for giving me company in early morning hours when I used to be the only person awake to write the book.

I am thankful to the whole team of CBS Publishers & Distributors, Mr SK Jain, CMD; Mr YN Arjuna, Senior Vice President—Publishing, Editorial and Publicity; Ms Ritu Chawla, General Manager; Mr Vikrant Sharma, Mr Neeraj Prasad, Mr Manish Raj, and Mr Prasenjit Paul for giving me this opportunity to publish my work.

Poonam Agrawal
drpoonam24agrawal@yahoo.com

Contents

Section III: Amino Acid Metabolism

Section IV: Enzymes

Section V: Vitamins

Section VI: Mineral Metabolism

Section VII: Nutrition

Section VIII: Heme Metabolism

Section IX: Hemoglobinopathy

Section X: Collagenopathy

Section XI: Nucleotide Metabolism

Section XII: Miscellaneous

CASE 1

Classical Galactosemia

A 4-month-old male child presents with the history of repeated loss of consciousness and refusal to feed specially milk and milk-containing diet. On examination, baby was found to be mild icteric and bilateral cataract* was detected. Liver was palpable below costal margin.

Following are the results of various laboratory investigations:

- Blood sugar: 72 mg/dl (normal random blood glucose = 80–140 mg/dl)
- Plasma-free galactose: 129 mg/dl (normal = <20 mg/dl)
- Serum uric acid: 8.4 mg/dl (normal = 2.5–7.2 mg/dl)
- Blood lactic acid: 4.8 mmol/L (normal = 0.5 to 1.0 mmol/L)
- RBC galactose-1-phosphate level: 54 mg/dl (normal = <1 mg/dl)
- RBC galactose-1-phosphate uridyltransferase activity was absent.

QUESTIONS

Q.1. What is the probable diagnosis in this case? What is the biochemical defect in this disease?

Q.2. Explain the biochemical reason for the above clinical signs and symptoms.

Q.3. What is the urinary finding in this case?

Q.4. What is the treatment regime suggested for this child?

Explanations

Ans.1. This child is suffering with **classical galactosemia** which is due to deficiency of enzyme: **'Galactose-1-phosphate uridyltransferase' (GaliPUT).** This enzyme is responsible for conversion of galactose-1-phosphate to glucose-1-phosphate. It is an autosomal recessive disorder. The pathway of galactose metabolism is illustrated in Fig. 1.

Ans.2. Biochemical explanation of clinical signs and symptoms is as follows:

a. **Hepatomegaly*:** Lack of this enzyme impairs galactose metabolism and results in accumulation of **galactose-1-phosphate** in liver which results in **hepatomegaly.**

b. **Recurrent hypoglycemic attack:** Accumulated galactose-1-phosphate inhibits glycogen phosphorylase enzyme which is manifested as **recurrent hypoglycemic**

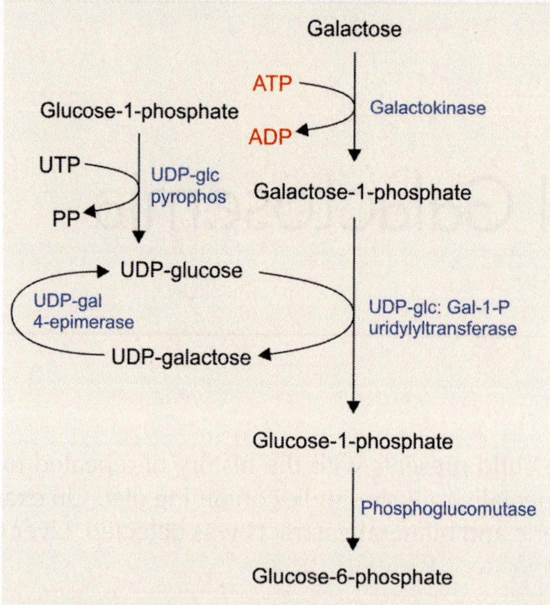

Fig. 1: Biochemical pathway involved in galactose metabolism

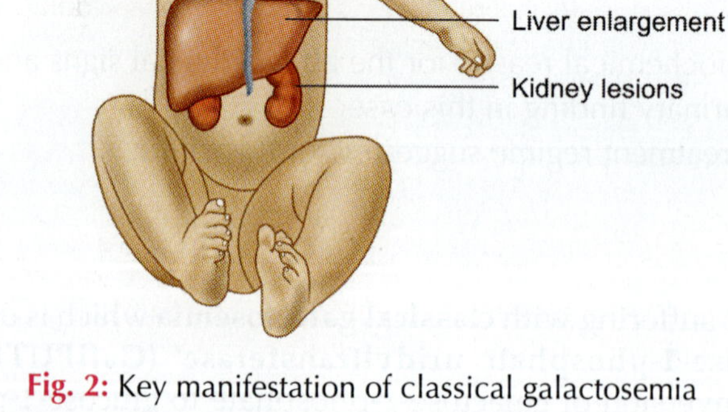

Fig. 2: Key manifestation of classical galactosemia

episodes. This is due to the fact that glycogen phosphorylase is the key enzyme in the liver to supply glucose in blood by glycogenolysis at the time of need. Once inhibited, this enzyme does not run the glycogenolysis and results in recurrent hypoglycemia.

c. **Cataract***: Cataract in the baby is due to reduction of unmetabolized galactose in the lens of the eye in **polyol pathway**, where **aldol reductase** converts galactose to galactitol which initiates cataract formation. Cataract is bilateral and in most of the cases, cataract is mild and transient and is reversible provided dietary treatment with lactose- and galactose-free diet, starts early within 20 days of life.

Ans.3. Galactose is not metabolized in galactosuria. This results in higher level of galactose circulating in blood which gets filtered out in the urine.

Benedicts test done in urine of such patients gives positive result as galactose is a reducing carbohydrate.

Ans.4. **Treatment of this child is lactose and galactose-free diet. This regime should be started within 10 days of life. Delay in starting the treatment results in organ damage and low IQ.** Despite adequate treatment from an early age, children with classic galactosemia remain at increased risk for developmental delays, speech problems and motor dysfunction.

Childhood **apraxia* of speech** and **dysarthria*** require expert **speech therapy.** Cataract may require surgery. If left untreated, galactosemic babies are prone to develop *E. coli* infection and may land up in sepsis and death.

Other important points related to this case:

- *Food items need to be avoided in galactosemic baby*: Breast milk, cow milk, casein and whey-containing food, medications having galactose and lactose to be avoided in galactosemia.
- Prenatal genetic diagnosis is possible in cultured amniotic fluid for **galactose-1-phosphate uridyltransferase enzyme assay.**

Bibliography

1. Casa RD, Ungaro C, Acampora E, *et al.* A case of galactosemia misdiagnosed as cow's milk intolerance. *Ital J Pediatr* 2012;38:47. https://doi.org/10.1186/1824-7288-38-47
2. Singh R, Kaur G, Thapa BR, Prasad R, Kulkarni K. A case of classical galactosemia: identification and characterization of 3 distinct mutations in galactose-1-phosphate uridyl transferase (GALT) gene in a single family. Indian J Pediatr. 2011 Jul;78(7):874–6. doi: 10.1007/s12098-010-0348-y. Epub 2010 Dec 28. Erratum in: Indian J Pediatr. 2011 Jul;78(7):894. Kulkarni, Ketan [added]. PMID: 21188552.

Word Meaning

Hepatomegaly: Enlargement of liver.

Cataract: A disease which causes dense and cloudy lens of eye which impairs vision.

Apraxia: Inability to perform certain task.

Dysarthria: Difficulty in articulation of speech.

CASE

2

Glucose-6-Phosphate Dehydrogenase Deficiency

A 23-year-old African boy is brought with fever, chills and rigors*; on examination, spleen was palpable, **antimalarial drug primaquine** was given. After two days, patient visit again with history of fatigue, breathlessness, pale skin, passing dark-colored urine. He also reported having abdominal discomfort.

Following are the results of various laboratory investigations:

- Hemoglobin: 8 g% (normal = 12–15 g%)
- Total bilirubin: 5.3 mg% (normal = 0–1.00 mg/dl)
- Conjugated bilirubin: 3.3 mg/dl (normal = 0–0.4 mg/dl)
- Unconjugated bilirubin: 2.0 mg/dl (normal = 0–0.6 mg/dl)
- SGPT: 65 IU/L (normal = 15–45 IU/L)
- SGOT: 148 IU/L (normal = 15–45 IU/L)
- Blood urea: 48 mg/dl (normal = 15–45 mg/dl)
- Serum creatinine: 0.82 mg/dl (normal = 0.6–1.2 mg/dl)

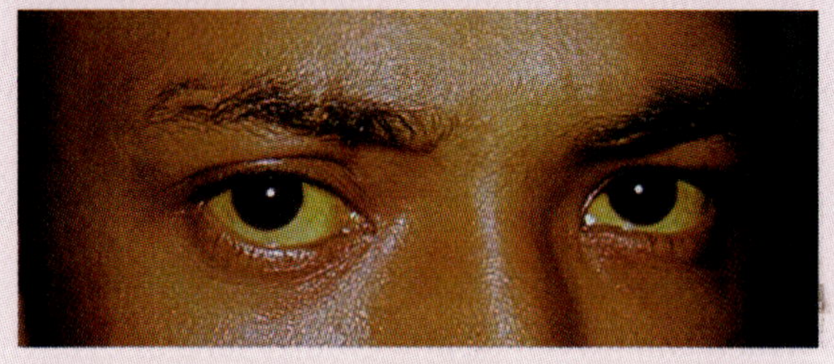

Jaundice with icteric sclera

QUESTIONS

Q.1. Explain the possible reason for such a presentation in this male.

Q.2. What is the biochemical basis of the signs and symptoms?

Q.3. Mention the factors which may precipitate the attack of **acute hemolytic anemia (AHA)** in patient having underlying G6PD deficiency. What Is **favism**?

Q.4. Why the deficiency of G6PD is more common in geographical areas where there is high prevalent of malaria?

Q.5. What is the treatment of this disorder?

Explanations

Ans.1. This patient is suffering with **glucose-6-phosphate dehydrogenase (G6PD) deficiency.** G6PD is protein having 516 amino acid and approximately 140 mutation is known in it. G6PD is the rate-limiting enzyme of HMP shunt pathway which is the major pathway responsible for production of NADPH. This results in inadequate amount of NADPH production in HMP shunt pathway.

Most of the cases are due to **sporadic mutations*** and in inherited conditions, it is inherited as **X-linked disease.** The *G6PD* gene is located on the long arm (q) of the X chromosome (Xq28).

Most common **enzyme deficiency associated with hemolytic anemia** is **G6PD deficiency** which is responsible for 96% cases of enzyme deficient hemolytic anemia. Less than 4% cases are due to deficiency of **pyruvate kinase** and less than 1% cases of enzyme deficient hemolytic anemia are due to **phosphoglucomutase** deficiency.

Ans.2. Intake of **primaquine** (an antimalarial drug) has precipitated an attack of hemolytic anemia in this patient who is presenting with breathlessness and dark colored urine.

This patient has underlying deficiency of glucose-6-phosphate dehydrogenase which has compromised the production of NADPH. Deficiency of NADPH does not maintain **glutathione in reduced state** which is necessary to protect the cell's hemoglobin and is important to maintain the membrane integrity. This protects the RBC against highly reactive oxygen radicals **(oxidative stress)**.

Ans.3. Precipitating factors for **acute hemolytic anemia** (AHA) in G6PD deficiency are:

a. Drugs
b. **Fava beans**
c. Certain infections.

Drugs

Hemolytic anemia episodes can result from exposure to certain drugs like **antimalarial, antipyretics and sulfa drugs.** Detail lists of drugs responsible for precipitation of acute episode of hemolysis in patient having G6PD are as follows: Primaquine, Sulfacetamide,

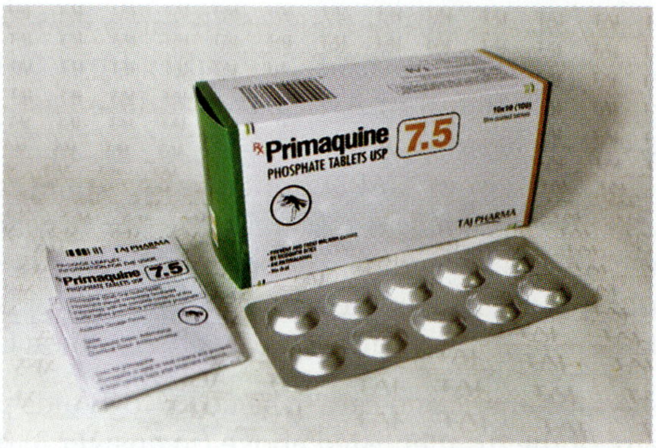

Fig. 1: Primaquine tablets precipitate hemolysis in G6PD deficiency

Norfloxacin, Acetanilid, Cotrimoxazole, Dapsone, Doxorubicin, Furazolidone, Methylene blue, Moxifloxacin, Nalidixic acid, Naphthalene, Niridazole, Nitrofurantoin, Pamaquine, Pentaquine, Phenazopyridine, Phenylhydrazine, Rasburicase, Sulfanilamide, Sulfapyridine, Thiazolesulfone, Toluidine Blue.

Fava Beans

Acute hemolytic anemia in G6PD-deficient people can develop after eating fava beans. This is known as **favism.** The chemicals, known as **vicine and convicine**, found within fava beans that trigger acute hemolytic anemia episodes in G6PD-deficient people.

Fig. 2: Fava beans

Typical Presentation of a Child having Hemolytic Episode

The symptoms begin 2 to 3 days after drug intake or infection but the time of onset of symptom is lesser in **favism.**

A child may have a slightly elevated temperature within 24–48 hours and can become irritable and unruly, or subdued and lethargic. Nausea, abdominal pain and diarrhea may develop. Urine may become noticeably dark and can appear red, brown or even black. Affected children may become pale and their resting heart rate may be high **(tachycardia).** Jaundice can also develop and the liver and spleen may become enlarged.

Ans.4. G6PD deficiency affects individuals of all races and ethnic backgrounds. Approximately 400 million people worldwide suffer from G6PD deficiency. **The highest prevalence rates are found in Africa, the Middle East, certain parts of the Mediterranean, and certain areas in Asia.**

The geographic distribution of G6PD deficiency correlates strongly with the distribution of malaria. It is proved in research that *G6PD* gene mutation conveys **protection from malaria** in these regions. The specific manner how **G6PD deficiency protects against malaria is** not fully understood but possibility is that this protective quality is linked to the inability of malaria to grow efficiently in G6PD-deficient cells.

Ans.5. Treatment is symptomatic and care to be taken to avoids precipitating factors. Before prescribing any of the drug enlisted above (in answer 3) drug history need to be explored and any previous episode of hemolysis need to be ruled out.

Bibliography

1. Cossio de Gurrola G, Araúz JJ, Durán E, et al. Kernicterus by glucose-6-phosphate dehydrogenase deficiency: a case report and review of the literature. J Med Case Reports 2, 146 (2008). https://doi.org/10.1186/1752-1947-2-146
2. La Vieille S, Lefebvre DE, Khalid AF, Decan MR, Godefroy S. Dietary restrictions for people with glucose-6-phosphate dehydrogenase deficiency. Nutr Rev. 2019 Feb 1;77(2):96–106. doi: 10.1093/nutrit/nuy053. PMID: 30380124.
3. Pinto A, MacDonald A, Cleto E, Almeida MF, Ramos PC, Rocha JC. A case report of a 4-year-old child with glucose-6-phosphate dehydrogenase deficiency: An evidence based approach to nutritional management. Turk J Pediatr. 2017;59(2):189–92. doi: 10.24953/turkjped.2017.02.012. PMID: 29276873.
4. Schuurman M, van Waardenburg D, Da Costa J, Niemarkt H, Leroy P. Severe hemolysis and methemoglobinemia following fava beans ingestion in glucose-6-phosphate dehydrogenase deficiency: case report and literature review. Eur J Pediatr. 2009 Jul;168(7):779–82. doi: 10.1007/s00431-009-0952-x. Epub 2009 Mar 5. PMID: 19263080.
5. Sköld MB, Svendsen RP, Pedersen EB. (Favism after ingestion of fava beans in a 3-year-old child with glucose-6-phosphate dehydrogenase deficiency.) UgeskrLaeger. 2017 May 15;179(20):V01170004. Danish. PMID: 28504635.

Word Meaning

Chills and rigors: Feeling of cold and shivering.

Sporadic mutations: Mutation which is not inherited from parents, rather is acquired.

Tachycardia: Rapid heart rate.

Palpable: Which can be felt by hand.

Oxidative stress: State of excess free radical/reactive oxygen species in the system.

Lethargic: Feeling of weak and sluggish, not interested in doing anything.

CASE

3

Hereditary Fructose Intolerance

A 3-year-old boy is brought in an unconscious state. He had headache, dizziness and a history of intake of **sugarcane juice** followed by weakness. On examination, hepatomegaly was noticed. Baby was icteric and knee joint tenderness was observed.

Following are the results of various laboratory investigations:

- Blood glucose: 40 mg/dl (normal random blood glucose = 80–140 mg/dl)
- Serum uric acid: 7.9 mg/dl (normal = 2.5–7.2 mg/dl)
- Blood urea: 87 mg/dl (normal = 15–45 mg/dl)
- Serum creatinine: 1.9 mg/dl (normal = 0.6–1.2 mg/dl)
- Blood lactate was high

QUESTIONS

Q.1. What is the most probable diagnosis in this case?

Q.2. What is the biochemical basis of the signs and symptoms?

Q.3. What is the treatment advised?

Q.4. What other disorder is associated with deranged fructose metabolism?

Explanations

Ans.1. This child is suffering with **hereditary fructose intolerance (HFI)** which is due to deficiency of **aldolase B**.

Hereditary fructose intolerance is an autosomal recessive disorder with a frequency of 1 in 20,000. The gene for human aldolase B has been mapped to chromosome 9q22.3.

A single missense "mutation, G to C transversion" is the most common mutation found in this disease.

Ans.2. Biochemical basis of signs and symptoms

Hepatomegaly

Aldolase B is the enzyme which is responsible for splitting **fructose-1-phosphate** into **dihydroxyacetone phosphate and glyceraldehyde.**

Out of three isoenzymes of aldolase in the liver, aldolase A, B and C, aldolase B is predominantly found in the liver.

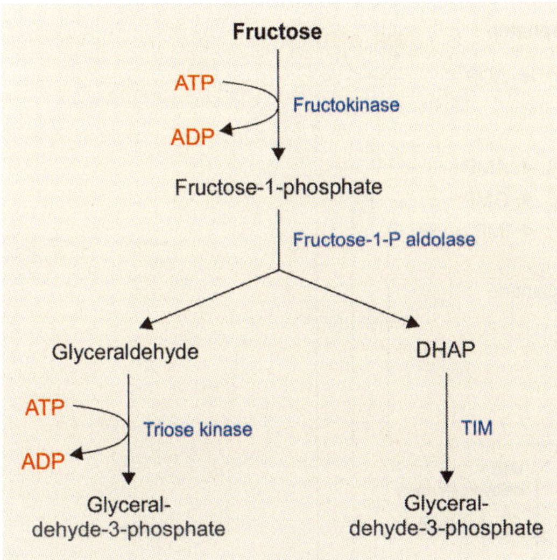

Fig. 1: Fructose metabolism in liver

In deficiency of aldolase B, fructose metabolism is affected and fructose-1-phosphate is not further metabolized. Deposition of fructose-1-phosphate results in hepatomegaly.

In addition, sequestration of Pi in the form of fructose-1-phosphate results in absolute deficiency of Pi in the cell which hampers production of ATP by oxidative phosphorylation. Reduced ATP affects ATP-dependent pumps which otherwise are responsible for maintaining the ionic gradient. This results in osmotic lysis of cells.

Hypoglycemia

Glycogenolysis is affected due to inhibition of glycogen phosphorylase by fructose-1-phosphate, resulting in hypoglycemic episode.

Lactic Acidosis

ATP is utilized in converting fructose to fructose-1-phosphate. This results in acute shortage of ATP which increases glycolysis. Increased glycolysis in limited oxygen supply in liver results in lactic acidosis **(Crabtree effect)**.

Hyperuricemia

ATP is a purine nucleotide and it imparts inhibitory affect on purine nucleotide biosynthesis by inhibiting **PRPP synthetase** enzyme.

Increased purine nucleotide biosynthesis (due to sequestration of ATP in converting fructose to fructose-1-phosphate) and then its degradation results in hyperuricemia. Impaired excretion of uric acid in lactic acidosis is also responsible for hyperuricemia. Many of such babies develop kidney failure.

Ans.3. Treatment of such a baby is complete removal of sucrose, fructose and sorbitol from diet which is certainly very difficult to implement because most fruits, vegetables and medicinal preparation contain fructose. Prognosis is poor in this condition and baby dies because of liver and kidney disorders.

Note: In a few case reports, **bilateral cataract** has been reported in babies having **hereditary fructose intolerance (HFI)**. This is said to be due to inhibitory effect of fructose on sorbitol dehydrogenase which results in excess accumulation of sorbitol

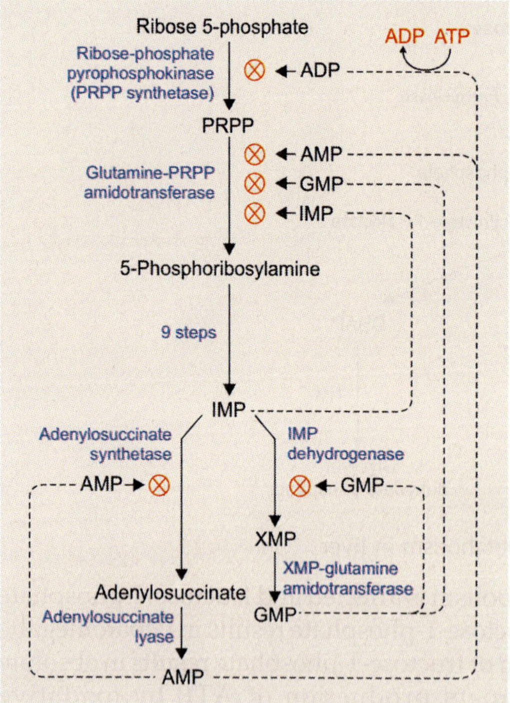

Fig. 2: Regulation of purine nucleotide biosynthesis. ATP is required for supply of ADP which inhibits action of PRPP synthetase

in lens cell. **Sorbitol** being a hygroscopic alcohol absorbs water and results in osmotic lysis of cell and initiation of cataract formation.

Ans.4. Other than hereditary fructose intolerance which is due to deficiency of aldolase B, there occurs deficiency of fructokinase enzyme which results in disorder known as benign fructosuria or essential fructosuria.

As the name itself implies, benign fructosuria is a condition which is a totally benign condition and not harmful. Affected children grow normally. It is detected accidently when urine shows positive Benedict test (due to excretion of fructose which is a reducing sugar) in absence of diabetes.

Bibliography

1. Ananth N, Praveenkumar GS, Rao KA, Vasanthi, Kakkilaya S. Two cases of hereditary fructose intolerance. Indian J Clin Biochem. 2003 Jul;18(2):87–92. doi: 10.1007/BF02867372. PMID: 23105397; PMCID: PMC3453872.
2. Eugênia Ribeiro Valadares, et al. Hereditary fructose intolerance in Brazilian patients. Molecular Genetics and Metabolism Reports. 2015; 4:35–38.
3. Ursula B. Kaiser, Robert A. Hegele. **Case Report: Heterogeneity of Aldolase B in Hereditary Fructose Intolerance. The American Journal of the Medical Sciences. 1991;302(6):364–68.**

Word Meaning

Missense mutation: A subtype of point mutation where change of nucleotide results in occurrence of a new codon which codes for amino acid which is different from original amino acid.

Transversion: A type of point mutation where purine nucleotide is changed for pyrimidine nucleotide and vice versa.

CASE

4

Lactose Intolerance

A 1-year-old boy presents to hospital with complain of swelling abdomen, pain and diarrhea after taking milk and other dairy products. These symptoms were existing for past 4 months and started soon after weaning was introduced. Baby is totally alright when fruit juice is given. On examination, perianal skin showed erythema* which may be due to frequent passage of stool with low pH.

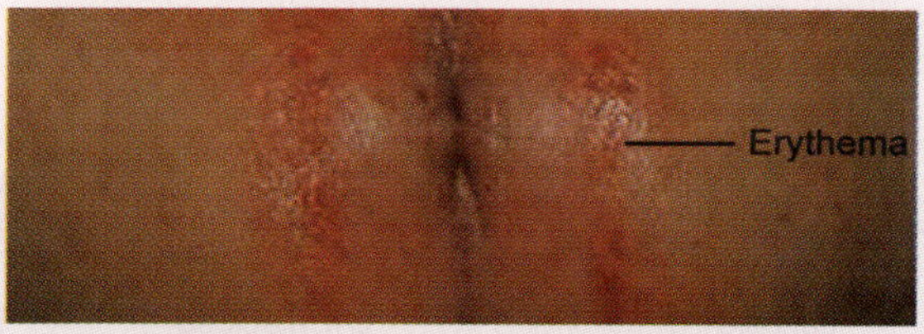

Perianal skin showing erythema due to frequent stool of low pH

QUESTIONS

Q.1. What is the most probable diagnosis in this case?
Q.2. What is the enzyme deficiency in this case?
Q.3. What is the biochemical basis of signs and symptoms in this disease?
Q.4. How to diagnose a case of lactose intolerance in a baby?
Q.5. What is the dietary advice given to mother?
Q.6. What are other clinical variants of this disorder? What is adult hypolactasia?

Explanations

Ans.1. Consumption of dairy products precipitate symptoms while on intake of fruit juice, baby is alright. This history suggests that boy is suffering with **lactose intolerance.**

Ans.2. This is due to deficiency of **lactase** enzyme (which is **beta galactosidase**). The gene of enzyme lactase is located on the long arm of chromosome 2 (region 2q21). Maximum lactase expression in intestinal cell occurs during the first months of life and declines after weaning.

Deficiency of this enzyme is common in preterm infant. This is the last **disaccharidase** which develops during intrauterine development, hence the deficiency of this enzyme is commonly seen in babies born prematurely.

The lactase enzyme is often situated in the microvilli of the small intestine (specially mid-jejunum) and it hydrolyses dietary lactose into its monomer D-glucose and D-galactose which are then being absorbed.

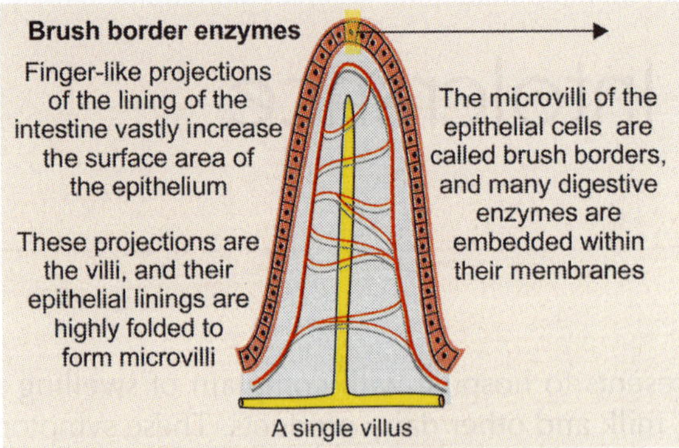

Fig. 1: Lactase is a brush border enzyme

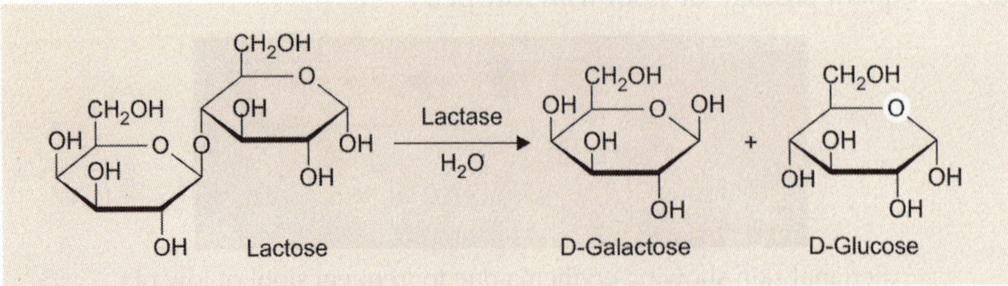

Fig. 2: Action of lactase on lactose

The enzyme spans the apical membrane of mature enterocytes and is homodimer (made up of two identical extracellular 160 kDa polypeptide chains) as well as a short intracytoplasmic part.

In many populations, **lactase levels** decline after weaning. This condition is called **lactase non-persistence (LNP).** LNP affects approximately 70% of the world's population and is the underlying physiological factor for primary lactose intolerance (LI).

While the decline in lactase levels starts soon after weaning, symptoms generally do not manifest before 5 years of age. Presentation in earlier age is most commonly associated with underlying gut condition like viral gastroenteritis, cow milk allergy, giardiasis, coelic disease, etc.

Ans.3.

BIOCHEMICAL BASIS OF SIGNS AND SYMPTOMS IN A CASE OF LACTOSE INTOLERANCE

Abdominal Distension and Bloating

The undigested and unabsorbed lactose in the lumen is osmotically active and tends to absorb fluid, resulting in diarrhea. It also becomes a substrate for action of intestinal bacteria which ferment the lactose in the colon to short-chain fatty acids (SCFA), hydrogen (H_2), carbon dioxide (CO_2) and methane (CH_4).

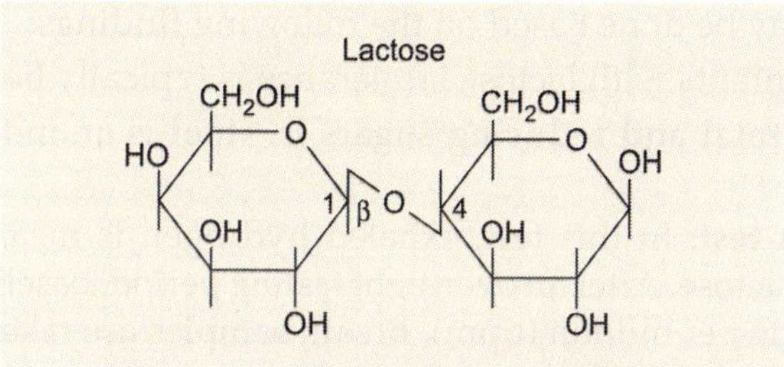

Fig. 3: Lactose is a disaccharide

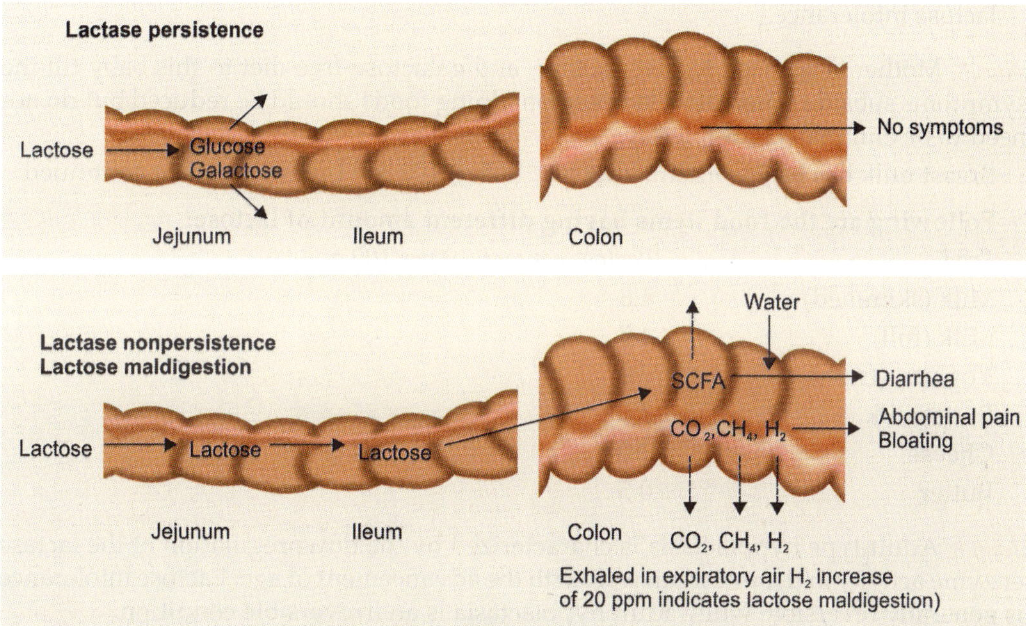

Fig. 4: Mechanism involved in abdominal pain, bloating in lactase nonpersistence (LNP)

These gases are responsible for flatulence, bloating and distension in the abdomen after consumption of milk and dairy product.

Fig. 5: Lactose-containing food items

Ans.4. Diagnosis may be done based on the following findings:

a. The **stool pH** in infants with lactose intolerance is typically **below 5.5 to 6.0.**
b. Measurement of **total and reducing sugars in stool** is an indirect test for lactose malabsorption.
c. **Breath hydrogen test:** In this test, exhaled hydrogen is measured after giving a standard dose of lactose. After an overnight fasting period, baseline hydrogen should be closed to 0 parts per million (ppm). Breath samples are taken every 30 min for 3 hr (from time of the lactose bolus).

   A rise in exhaled hydrogen by 0.20 ppm from baseline is considered diagnostic of lactose intolerance.

Ans.5. Mother is advised to give lactose- and galactose-free diet to this baby till the symptoms subside. Thereafter lactose-containing foods should be reduced but do not need to be eliminated completely.

Breast milk has high content of lactose (7.5 g/100 ml) but it should be continued.

Following are the food items having different amount of lactose:

Food	*Lactose content (g) per 100 g*
Milk (skimmed)	4.8
Milk (full)	4.7
Yoghurt (fresh)	3.0
Buttermilk	3.0
Cheese	3.0
Butter	0.5

Ans.6. Adult-type hypolactasia is characterized by the downregulation of the lactase enzyme activity in the intestinal wall with the advancement of age. Lactose intolerance is generally reversible while adult hypolactasia is an irreversible condition.

Bibliography

1. Järvelä IE. Molecular genetics of adult-type hypolactasia. Ann Med. 2005;37(3):179–85. doi: 10.1080/07853890510007359. PMID: 16019716.
2. Misselwitz B, Butter M, Verbeke K, Fox MR. Update on lactose malabsorption and intolerance: pathogenesis, diagnosis and clinical management. Gut. 2019 Nov;68(11):2080–91. doi: 10.1136/gutjnl-2019-318404. Epub 2019 Aug 19. PMID: 31427404; PMCID: PMC6839734.

Word Meaning

Erythema: Redness of the skin.

CASE

5

Sucrose Intolerance

A 5-week-old male child is admitted with history of diarrhea for past one week, irritability and refusal to feed for past 4 days. On careful exploring the history, it became evident that baby was exclusively breastfed for 3 weeks after which he was given dried milk powder with added sugar to enhance the taste. He developed diarrhea along with vomiting when dried milk powder was introduced with added sugar.

On examination, baby was dehydrated and irritable and was continuously screaming. Abdomen was full and tender.

Routine blood investigation for complete blood count (CBC), ESR was normal. Blood biochemistry showed normal results for various parameters. Stool culture was negative for bacteria and parasites. Baby was admitted and was treated.

Baby was found to have aversion to sweet food. Symptoms subsided when sucrose was replaced by glucose. He improved and was discharged after 4 days of admission with advice for treatment for dehydration and sugar-free diet.

QUESTIONS

Q.1. What is the diagnosis in this case?

Q.2. Explain the biochemical basis of signs and symptoms in this baby.

Q.3. How to confirm the diagnosis?

Q.4. What is the dietary advice given to mother of this child?

Q.5. What is congenital sucrase isomaltase deficiency (CSID)? What is the gold standard for diagnosis of CSID?

Q.6. How will you detect the presence of sucrose in the urine?

Explanations

Ans.1. This baby is having **sucrose intolerance.** In this condition, there is deficiency of **sucrase enzyme** which is required to cleave sucrose into glucose and fructose.

Ans.2. Deficiency of enzyme sucrase results in no digestion of sucrose which then is acted upon by intestinal bacteria which convert sucrose to various gases. This gas in the intestinal lumen results in bloating, distension and abdominal pain. Baby slowly learns to avoid sweet food and develops aversion.

Ans.3. Noninvasive test is **sucrose tolerance test and ^{13}C sucrose breath hydrogen test.**

Sucrose tolerance test is done by giving 2 g/kg of sucrose as 20% solution and blood glucose estimation is done in 30 min, 60 min, 120 min and 180 min time and is compared with fasting. Flat curve is seen in patients suffering with sucrose intolerance along with watery diarrhea after instituting this test.

This test is considered normal when there is rise of at least 20 mg/dl in glucose level compared to fasting state during the test and absence of diarrhea within 4 hours of test.

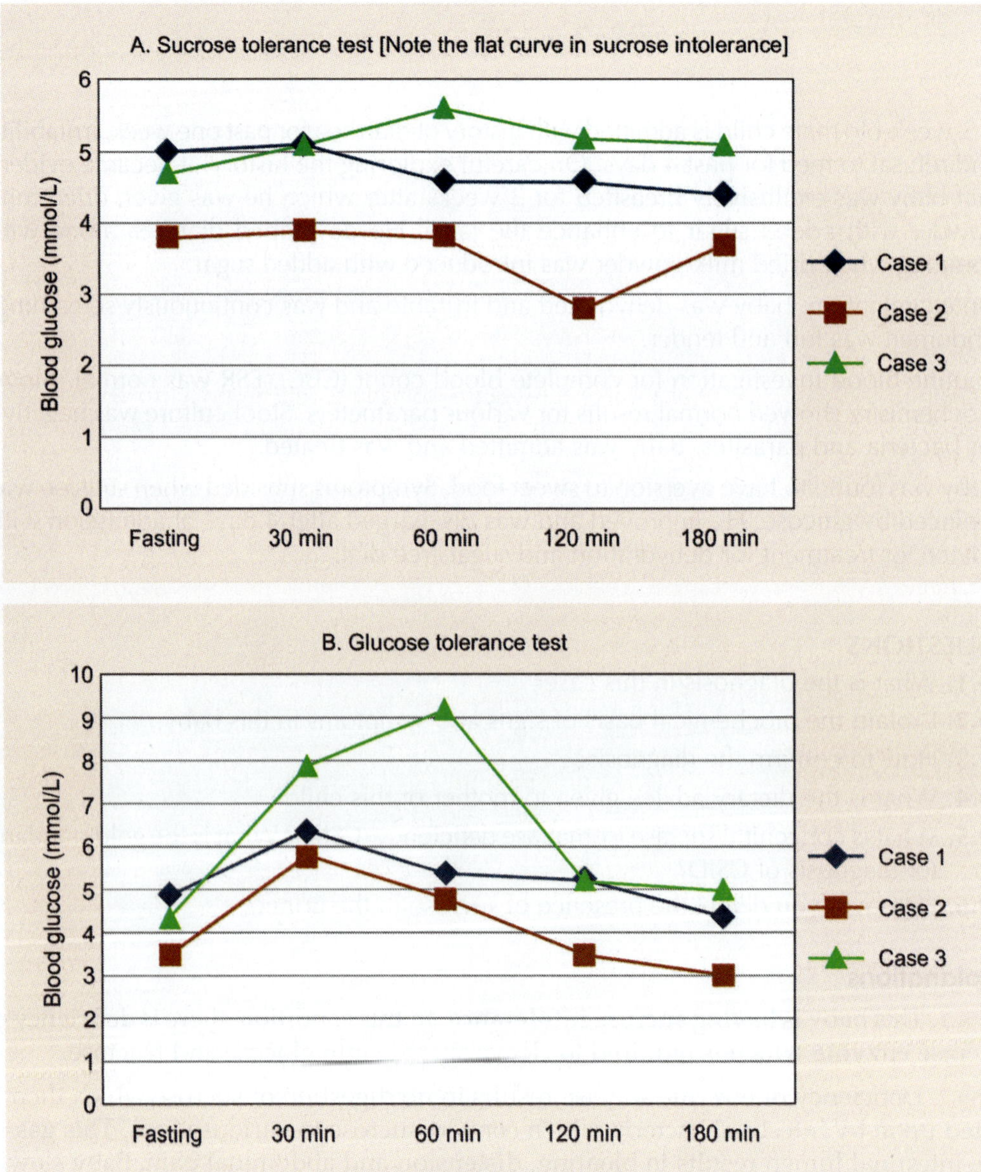

Fig. 1: Sucrose and glucose tolerance tests

Ans.4. Sucrose (table sugar) free diet is advised lifelong.

Sucrase (sacrosidase) has been shown to be highly effective, leading to relief of symptoms and improved nutritional status in certain studies.

Baker's yeast consists of lyophilized *Saccharomyces cerevisiae* and has enough sucrase activity, mild isomaltase activity and no lactase activity. It can be advised in sucrose intolerance patient.

Hereditary sucrose intolerance in adult precipitated by gastroenteritis is rarely reported in literatures.

Ans.5. Congenital sucrase isomaltase deficiency (CSID) is combined sucrase and isomaltase deficiency. It is an autosomal recessive (AR) disorder. Gold standard for diagnosis of CSID is small intestine biopsy for sucrase-isomaltase **(disaccharidase)** activity.

Ans.6. Sucrose can be detected in the urine by sucrose hydrolysis test. In this test first step is to do Benedict's test with the urine containing sucrose. This will result in negative test result as the sucrose is the nonreducing sugar. Thereafter, two drops of dilute HCl is added in the urine and it is boiled. After cooling, Benedict test is performed once again with such urine which now will give positive response as sucrose after acid hydrolysis yields glucose and fructose which are reducing sugars.

Bibliography

1. Geng L, Li DY, Ou W, *et al.* Congenital sucrase-isomaltase deficiency: an under-diagnosed disease in Chinese children. *BMC Pediatr* 14, 11 (2014). https://doi.org/10.1186/1471-2431-14–11.
2. Neale G, Clark M, Levin B. Intestinal Sucrase Deficiency Presenting As Sucrose Intolerance In Adult Life. *The British Medical Journal,* 1965;2(5472):1223–25. Retrieved November 15, 2020, from http://www.jstor.org/stable/25404513
3. Zimmer Klaus-Peter, Scholz Dietmar, Naim Hassan Y. Transient Sucrose and Starch Intolerance, Journal of Pediatric Gastroenterology and Nutrition: November 2012;55:S39–S40. doi: 10.1097/01.mpg.0000421411.10999.ad

CASE

6

Diabetes Mellitus

A 56-year-old lady comes with complain of increased thirst (polydipsia), increased appetite (polyphagia) and increased frequency of urine (polyuria) for past 6 months.

Following are the results of various laboratory investigations:

- Fasting plasma glucose: 146 mg/dl (normal = 80–110 mg/dl)
- Postprandial blood sugar: 209 mg/dl (normal = <140 mg/dl)
- HbA1c: 7.1% (normal = <6.5%)
- Urine is positive for sugar in dipstick test.

QUESTIONS

Q.1. What is the condition patient is suffering from?

Q.2. What are American Diabetic Association (ADA) and WHO guidelines for diagnosing the diabetes?

Q.3. What are the criteria currently adopted to diagnose diabetes mellitus?

Q.4. In addition to blood sugar and HbA1c, what other investigations would you like to do in this patient?

Q.5. What is the reason behind **polyphagia** and **polyuria** in this condition?

Q.6. What is the line of treatment?

Q.7. What is HbA1c? What is the most common method of HbA1c estimation?

Q.8. Name short-term and long-term complications of diabetes mellitus.

Q.9. What is the conversion factor for converting mg/dl glucose values to mmol/L?

Q.10. What do you understand by the term gestational diabetes mellitus (GDM)?

Explanations

Ans.1. This patient is suffering with **diabetes mellitus**. Diabetes mellitus is the disease which is due to lack of insulin action on various metabolic pathways. This may be either due to absolute lack of insulin or may occur because of insulin resistance.

Ans.2. The ADA and WHO guidelines for diagnosing diabetes are given in Tables 1 and 2.

Table 1: American Diabetes Association (ADA) diagnostic criteria for diabetes

Test	*Threshold*	*Qualifier*
Hemoglobin A1c or	≥6.5%	Lab NGSP-certified, standardized DCCT assay
Fasting glucose or	≥126 mg/dl (7.0 mmol/L)	No caloric intake for at least 8 hours
2-hour glucose or	≥200 mg/dl (11.1 mmol/L)	After 75 g of anhydrous glucose
Random glucose	≥200 mg/dl (11.1 mmol/L)	Plus classic hyperglycemia symptoms or crisis

NGSP, National Glycohemoglobin Standardization Program; DCCT, Diabetes Control and Complications Trial.
*Results must be confirmed by repeated testing.

Table 2: WHO diagnostic criteria for diabetes

Condition unit	*2-hour glucose mmol/L (mg/dl)*	*Fasting glucose mmol/L (mg/dl)*	*HbA1c*	
			mmol/mol	*DCCT%*
Normal	<7.8 (<140)	<6.1 (<110)	<42	<6.0
Impaired fasting glycemia	<7.8 (<140)	≥6.1 (≥110) and <7.0 (<126)	42–46	6.0–6.4
Impaired glucose tolerance	≥7.8 (≥140)	<7.0 (<126)	42–46	6.0–6.4
Diabetes mellitus	≥11.1 (≥200)	≥7.0 (≥ 126)	≥48	≥6.5

Ans.3. The international expert committee with the members appointed by American Diabetes Association (ADA), **European Association for the Study of Diabetes,** and the IDF have issued the following **diagnostic criteria for diagnosis of diabetes:**

a. HbA1c ≥6.5%, **or**

b. FPG ≥126 mg/dl, **or**

c. Two-hour postprandial* plasma glucose ≥200 mg/dl after oral glucose tolerance test (OGTT) (75 g of anhydrous glucose dissolved in water), **or**

d. Random blood glucose ≥200 mg/dl along with the symptoms of diabetes mellitus.

Ans.4. In addition to blood sugar and HbA1c, lipid profile, urine **for microalbumin** should be importantly advised.

Patient should be thoroughly worked out to rule out diabetic complication.

Urine should be investigated for microalbuminuria to rule out impending danger of **diabetic nephropathy*.**

Lipid profile should be done, as diabetic patients are prone for **dyslipidemias*.**

Ans.5. In diabetes mellitus, due to lack of insulin, glucose is not adequately metabolized and fails to produce required ATP. Patient thus constantly feels hungry and that is the reason behind **polyphagia*.**

Polyuria* in diabetes is **osmotic diuresis*** and results due to increased excretion of glucose in the urine which drags water along with it.

Ans.6. Patient should be advised **oral hypoglycemic treatment** along with dietary advice and **physical exercise** to enhance **insulin sensitivity.** Frequent monitoring of blood glucose is to be done and antidiabetic medication dose should be adjusted accordingly.

Ans.7. HbA1c is known as glycosylated hemoglobin. It is formed due to nonenzymatic combination of glucose with amino terminal group of β chain of hemoglobin.

Initially the aldehyde group of glucose form a 'Schiff base linkage' with amino group which then is spontaneously changed to more stable 'amino ketone linkage'. This reaction is known as Amadori rearrangement. **Unit of HbA1c is % as it is the percentage of total hemoglobin which is undergoing such glycosylation. Concentration of HbA1c depends upon the status of glucose in the blood and duration of hyperglycemia.**

Normal value of HbA1c = <6.5%

Value of HbA1c ≥ 6.5% is a diagnostic criterion of DM. Assessment of level of HbA1c is important for diagnosis of diabetes mellitus and also to interpret effectiveness and compliance to diabetes treatment (Tables 3 and 4).

Table 3: HbA1c and diabetic status

HbA1c	*Status*
<5.7%	Non-diabetic
5.7–6.4%	Prediabetic
≥6.5%	Diabetes
<7%	Goal of therapy
>8%	Action suggested

Table 4: HbA1c and corresponding mean plasma [*Rule of 8]

HbA1c	*Mean plasma glucose (mg%)*
5	75
6	110
7	145
8	**180 mg%**
9	215

*Rule of 8: According to this, for HbA1c value of 8%, mean plasma value is taken as 180 mg% and for every 1% increase or decrease of HbA1c, 35 mg% of glucose is added or subtracted respectively from 180 mg% value to get the mean plasma glucose value.

Ans.8. DM is a metaolic disease characterized not only by its impact on various metabolic pathways, but also it is prone for developing various complications. Complications of DM can be classified as follows:

- **Short-term complications:** Hypoglycemia, diabetic ketoacidosis, hyperosmolar hyperosmotic coma
- **Long-term complications:** Retinopathy, nephropathy, neuropathy.

Ans.9. mg/dl value of glucose can be converted to mmol/L by dividing it with factor 18:

$$\text{mmol/L value of glucose} = \frac{\text{mg/dl value of glucose}}{18}$$

Similar fashion, the glucose values given in mmol/L can be converted to mg/dl value by multiplying it with factor 18.

$$\text{Value of glucose in mg/dl} = \text{Value of glucose in mmol/L} \times 18$$

Ans.10. In 2013, the World Health Organization (WHO) recommended that hyperglycemia first detected during pregnancy be classified as either 'diabetes mellitus (DM) in pregnancy' or 'GDM'.

Pregnancy is associated with insulin resistance (IR) and hyperinsulinemia that may predispose some women to develop diabetes. Gestational diabetes has been defined as any degree of glucose intolerance with an onset, or first recognition during pregnancy.

Bibliography

1. American Diabetes Association. Diagnosis and classification of diabetes mellitus. Diabetes Care. 2010 Jan;33 Suppl 1(Suppl 1):S62-9. doi: 10.2337/dc10-S062. Erratum in: Diabetes Care. 2010 Apr;33(4):e57. PMID: 20042775; PMCID: PMC2797383.

Word Meaning

Diabetic nephropathy: Kidney damage due to diabetes.

Dyslipidemia: Altered lipid parameters.

Polyphagia: Excessive eating

Polyuria: Excessive urination

Osmotic diuresis: Excessive urination due to filtered glucose in urine.

Postprandial: 2 hours after taking meal.

CASE

7

Diabetic Ketoacidosis

A 15-year-old boy is brought to emergency unit in a comatose state. He is a known diabetic since 5 years on irregular insulin treatment. Previous year he had an episode of loss of consciousness while playing for which he was treated with glucose and he recovered. On examination, his palm was cold and clammy and his breath was having fruity/sweet odor. Pulse rate was 120/min and respiration was fast and shallow.

Following are the results of various laboratory investigations:

- Blood glucose (random): 408 mg/dl (normal = random blood glucose = 80–140 mg/dl)
- pH = 7.05 (normal pH = 7.35–7.45)
- *Urine*
  - Protein—nil
  - Ketone bodies ++++
  - Sugar ++++

QUESTIONS

Q.1. What is the probable diagnosis?

Q.2. What is the biochemical basis of signs and symptoms?

Q.3. How is the process of ketogenesis regulated?

Q.4. What test will you do to find out presence of ketone body in the urine? What is the sensitivity and specificity of this test in diagnosing ketoacidosis?

Q.5. How is the ATP derived from ketolysis?

Explanations

Ans.1. This boy is having **diabetic ketoacidosis** (DKA) characterized by ketone body in body fluid (urine) and in breath. DKA is the complication of diabetes mellitus. Lack of insulin results in excessive mobilization of fatty acid and its oxidation results in ketoacidosis.

Ans.2. **Fruity odor of the breath** is due to loss of **acetone** in the breath. **Acetone** is one of the ketone bodies formed due to spontaneous decarboxylation of **acetoacetate.** Body has no mechanism to metabolize the acetone and being volatile this compound is exhaled in breath of the patient giving sweet **fruity odor to the breath.**

Ans.3.

Regulation of Ketogenesis

Ketogenesis is regulated at three important steps:

a. Level of fatty acid in plasma
b. Control at the level of CPT-I
c. Fate of acetyl-CoA

a. Level of Fatty Acid in Plasma

Precursor of ketone body is fatty acid. Approximately one-third of total content of fatty acid in the plasma is utilized by liver both in fed and starved states.

In an event of excessive mobilization of fatty acid from adipose cell during starvation, liver uses this fatty acid for ketone body production in addition of its other uses.

b. Control at the Level of CPT-I (Carnitine Palmitoyl Transferase I)

The major fraction of fatty acid taken up by the liver from circulation is oxidized in beta oxidation system and remainder is esterified to generate phospholipid and triacylglycerol.

Important step of regulation for diversion of fatty acid for either beta oxidation or esterification is **CPT-I enzyme** which is found at outer mitochondrial membrane.

CPT-I activity is low in well fed state. This is due to inhibitory action of malonyl-CoA (produced during well fed state) on CPT-I. In starvation and diabetes, malonyl-CoA is not synthesized hence the CPT-I is not inhibited.

This explains excess entry of fatty acid in mitochondrial matrix for beta oxidation during starvation and diabetes.

c. Fate of Acetyl-CoA

Diversion of acetyl-CoA to either TCA cycle or ketone body production is determined by the fact that ATP generated by acetyl-CoA should remain constant. Surplus acetyl-CoA is diverted to ketone body production.

Lack of oxaloacetate in TCA cycle due to its diversion in gluconeogenesis is also an important factor which determines deviation of acetyl-CoA for ketogenesis.

Ans.4. **Rothera's test** can be done in the urine to find out ketone body (acetone and acetoacetic acid).

This test is done in urine. 5 ml of urine is taken in test tube and is saturated with ammonium sulfate.

Freshly prepared **2% sodium nitroprusside** is mixed in urine. 1 ml of ammonia is added from the side of the test tube and purple ring formed at the junction of two liquids is observed. Appearance of ring suggests presence of ketone body in the urine.

Rothera's test is specific but not a very sensitive test to detect ketone body in the urine. Positive Rothera's test rules in the diagnosis of ketogenesis and ketonuria but negative Rothera's test does not rule out ketogenesis and associated ketonuria.

This is due to the fact that Rothera's test is positive for acetoacetate but not given by beta hydroxybutyrate, which incidently is the major ketone body in plasma and urine.

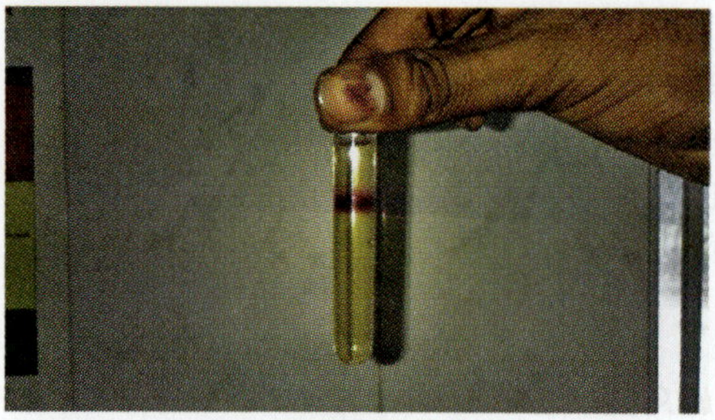

Fig. 1: Rothera's test done in urine showing positive test (note the purple color ring)

Ans.5. Ketolysis takes place in all the cells except in liver and RBC. Liver lacks thiophorase and RBC lacks mitochondria. It generates ATP in the following steps:

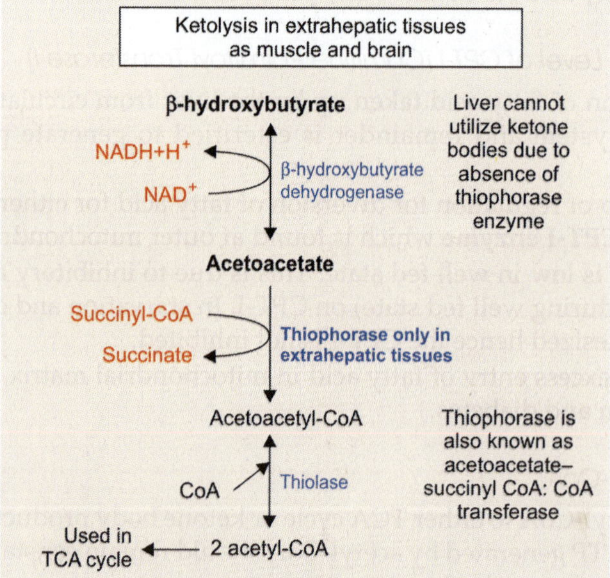

Fig. 2: Steps involved in ketolysis

i. Enzyme **beta hydroxybutyrate dehydrogenase** converts β-hydroxybutyrate to acetoacetate, with simultaneous production of NADH.
ii. Acetoacetate is converted to acetoacetyl-CoA by enzyme **thiophorase** which adds CoA moiety from succinyl-CoA.
iii. Acetoacetyl-CoA is cleaved to acetyl-CoA by enzyme thiolase.
iv. Acetyl-CoA is finally utilized in TCA cycle for production of ATP.

Bibliography

1. Comstock JP, Garber AJ. Ketonuria. In: Walker HK, Hall WD, Hurst JW, editors. Clinical Methods: The History, Physical, and Laboratory Examinations. 3rd edition. Boston: Butterworths; 1990. Chapter 140. Available from: https://www.ncbi.nlm.nih.gov/books/NBK247/
2. Kumar V, Nanavati SM, Melki G, Upadhyaya M, Dhillon R, Gibiezaite S, Michael P, Singhal M. A Case of Diabetic Ketoacidosis Presenting with Hypernatremia, Hyperosmolarity, and Altered Sensorium. Case Rep Endocrinol. 2018 Jul 16;2018:4806598. doi: 10.1155/2018/4806598. PMID: 30112226; PMCID: PMC6077649.

CASE

8

Correct Vacutainer for Blood Collection for Glucose Estimation

A 45-year-old male is admitted for routine workup. Attending intern collected 5 ml blood in plain vacutainer at 8 am for various organ function tests including blood sugar. Time of sample processing was 11 am and report was received at 12 noon. All laboratory results were normal except blood glucose which was 47 mg/dl.
Consultant biochemistry requested for repeat blood sample in correct vacutainer?

QUESTIONS

Q.1. What is the reason of low blood glucose values in blood collected in plain vacutainer?
Q.2. What is the rate of loss of blood glucose, if blood is collected in plain vacutainer?
Q.3. What all types of vacutainers are being used in various investigations? What vacutainer is recommended for collection of blood for blood glucose estimation and why?
Q.4. What is the preferred sample for blood glucose estimation?
Q.5. Mention various biochemical methods for glucose estimation. Which method is commonly used in routine diagnostic labs?
Q.6. Explain the mechanism by which fluoride inhibits enolase action?
Q.7. As a POCT (point of care testing) glucose is estimated using glucometer. Explain.

Explanations

Ans.1. The reason of low glucose in blood collected in plain vacutainer is spontaneous glycolysis which lowers blood glucose. Blood should be collected in **fluoride**

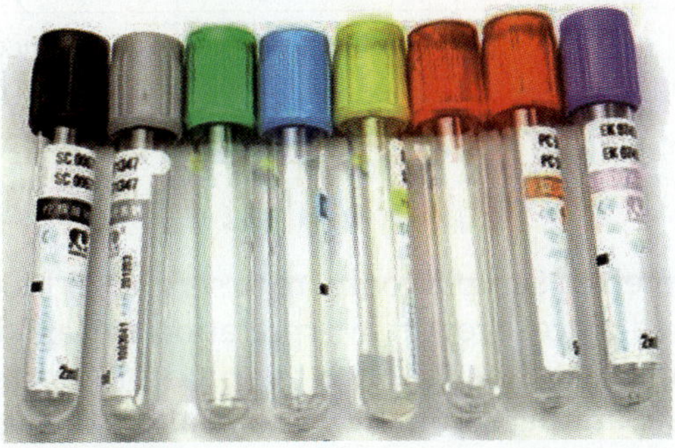

Fig. 1: Various types of vacutainers color coded

vacutainers for estimation of blood glucose as **sodium fluoride** acts as **antiglycolytic agent** and prevents loss of glucose by spontaneous glycolysis in collected blood.

Ans.2. Rate of loss of blood glucose is 10 mg/dl per hour, if blood is collected in plain vacutainer. Three-hour gap since the time of collection and actual analysis makes loss of 30 mg/dl in the above case.

Ans.3. Various types of vacutainers used in laboratory and their purpose are given in Table 1.

Table 1: Various types of vacutainers as coded by color of their cap and their purpose

Type of vacutainer/ color of the cap	*Representation*	*Additive*	*Purpose*	*Mixing instruction*
Red (glass) (clotting time: 60 min)		Nil	Serum biochemistry	Do not invert
Red (plastic) (clotting time: 60 min)		Silicone coating	Serum biochemistry	Invert 4 times
Yellow (clotting time: 30 min)		Clot activator	Serum biochemistry	Invert 4 times
Orange (clotting time: 5 min)		Rapid serum tube (thrombin based clot activator)	Serum biochemistry	Invert 4 times
Light blue		Sodium citrate (1:9)	Coagulation studies	Invert 4 times
Black		Sodium citrate (1:4)	Coagulation study	
Green		Heparin	ABG analysis	Invert 8 times
Lavender/purple		EDTA	Hemogram, ESR, HbA1c	Invert 8 times
Gray		Potassium oxalate and sodium fluoride	Blood glucose, blood alcohol	Invert 8 times

Ans.4. Plasma is the preferred sample compare to serum for blood glucose estimation as all the criteria for blood glucose report interpretation are laid down based on plasma values. Moreover, serum is extracted once the blood is clotted which takes a few minutes. This delay in separating serum from blood alters the value of glucose. Hence, plasma is preferred over serum for blood glucose result.

Ans.5. Various biochemical methods for glucose estimation are:

1. Glucose oxidase-peroxidase (GOD-POD) method.
2. King-Asatoor method.
3. Folin-Wu method.
4. Hexokinase method.

Most common method used in routine diagnostic lab is glucose oxidase-peroxidase (GOD-POD) method.

Ans.6. Enolase, also known as 2-phospho-D-glycerate hydrolase, is the enzyme of glycolysis (Embden-Meyerhof-Parnas) pathway. This enzyme catalyses reversible dehydration of 2-phospho-D-glycerate to produce PEP (phosphoenolpyruvate). This enzyme is dimer having two subunits of enzyme. Each of these two subunits require Mg^{++} for imparting the catalytic activity to this enzyme.

Fluoride interferes with accurate binding of magnesium ion to the enolase enzyme subunit and thus interfere with its catalytic activity.

Ans.7. POCT (Point-of-Care Testing) is also known as near patient testing or bedside testing is the testing system where testing is done near the patient at the point where patient care is being given. POCT is performed outside the laboratory and generally it gives the immediate result which enhances accurate patient care. Limited number of investigations are done in POCT. Glucose assessment by glucometer is one of them.

Bibliography

1. https://www.sciencedirect.com/topics/nursing-and-health-professions/blood-collection-tube
2. https://www.pathlabs.org/healthcare-providers/clinical-testing/clinical-tube-types.aspx

CASE

9

von Gierke Disease (Glycogen Storage Disorder Type Ia)

A 3-year-old boy presents with history of yellow discoloration of eyes and passing high colored urine for past 20 days. Mother gives the history of many episodes of loss of consciousness in the past. On examination, baby was small for age with round face and fatty cheeks and thin extremities. He was having swollen and tender right knee joint. On palpation, liver was found to be enlarged 7 cm below the costal margin. Baby was born by normal vaginal delivery at full term and his growth was normal till age of 1 year.

Following are the results of various laboratory investigations:

- Serum AST: 576 IU/L (normal = 15–45 IU/L)
- Serum ALT: 657 IU/L (normal = 15–45 IU/L)
- Serum ALP: 243 IU/L (normal = 98–278 IU/L)
- Total protein: 7.5 g/dl (normal = 5.5–8.5 g/dl)
- Albumin: 4.1 g/dl (normal = 3.5–5.5 g/dL
- Globulin: 3.4 g/dl (normal = 1.5–3.0 g/dl)

Random blood glucose

- Blood glucose: 47 mg/dl (normal random glucose = 80–140 mg/dl)
- Blood lactate: 6.6 mmol/L (normal lactate = 0.5–1 mmol/L)
- Serum uric acid: 7.9 mg/dl (normal = 2.5–7.0 mg/dl)
- pH: 7.31 (normal = 7.35 to 7.45)
- Ketone bodies in urine: +++

QUESTIONS

Q.1. What is the most probable diagnosis?

Q.2. What enzyme is deficient in this case and how many subtypes of type 1 GSD you know?

Q.3. Explain the biochemical basis of above signs and symptoms.

Q.4. How to approach to diagnosis?

Q.5. What is the treatment advised?

Explanations

Ans. 1 and 2. This baby is suffering with **von Gierke disease (type 1a glycogen storage disorder)** which is due to deficiency of endoplasmic reticulum enzyme **'glucose 6-phosphatase'** enzyme in **liver.** This enzyme is required at terminal step of glycogenolysis and gluconeogenesis and is responsible for conversion of glucose 6-phosphate to free glucose.

Type 1b is due to mutation of glucose 6-phosphate translocase on endoplasmic reticulum membrane. Gene map locus 17q21 is involved in type 1a and gene map locus 11q23 is involved in type 1b. **Type 1 (both 1a and 1b) is inherited as autosomal recessive** disorder.

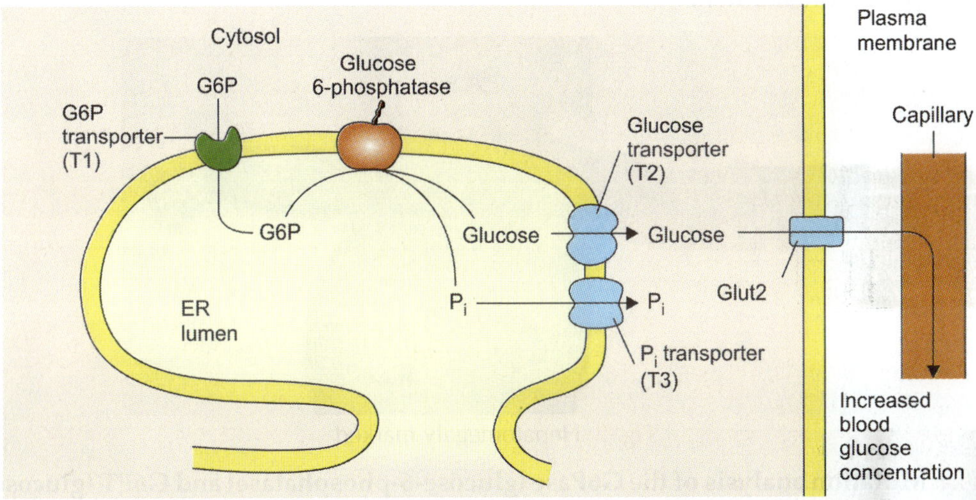

Fig. 1: Schematic plan of conversion of glucose 6-phosphate to free glucose

Type 1a constitutes 80% cases of type 1 and type 1b constitutes 20% of all cases of type 1. Annual worldwide incidence of type 1a GSD is 1 in 1,00,000 live births.

Ans.3.

Biochemical Explanation of Various Signs and Symptoms

Repeated Loss of Consciousness

von Gierke disease results in profound **hypoglycemia** and **ketoacidosis.** Hypoglycemia usually manifests first as tremors, seizures, cyanosis and apnea and, in the long-term, evolves to growth retardation.

High Blood Lactate Level

Lactate is high due to excess utilization of glucose 6-phosphate in hepatic cells for glycolysis in relative oxygen deficiency which results in **anaerobic glycolysis (Crabtree effect).**

Hyperuricemia

Hyperuricemia is due to excess purine nucleotide biosynthesis and then its catabolism resulting in high level of production of uric acid.

Excess synthesis of purine nucleotide is because of the following reasons:

a. **Excess production of ribose-5-phosphate in hexose monophosphate (HMP) shunt** pathway which is the precursor molecule for purine nucleotide biosynthesis.
b. Impaired excretion of uric acid due to **lactic acidosis** also compound to hyperuricemia.

In a nutshell, both overproduction and impaired excretion of uric acid are responsible for **hyperuricemia and gouty arthritis** in such cases.

Hepatomegaly

Liver enlargement is due to excessive deposition of glycogen due to normal synthesis and impaired breakdown.

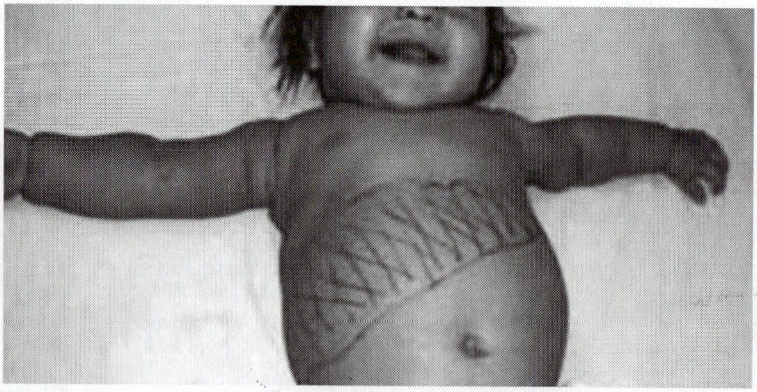

Fig. 2: Hepatomegaly marked

Ans.4. **Mutation analysis of the G6Pase (glucose-6-phosphatase) and G6PT (glucose-6-phosphate translocase) genes** by PCR-RFLP (restriction fragment length polymorphism), or by direct gene sequencing is done as a method of choice for diagnosis. In some cases, liver biopsy may be done for assessment of enzyme.

Ans.5. **Frequent feeding with glucose or corn starch** is advisable to **sustain the blood glucose level** so that baby does not get frequent attack of hypoglycemia.

Bibliography

1. Maurice M. Fliess, Samuel M. Bloom. von Gierke's glycogen storage disease: Case report with unusual findings, The Journal of Pediatrics, 1938;12(5):635–40, ISSN 0022-3476.
2. Raza M, Arif F, Giyanwani PR, Azizullah S, Kumari S. Dietary Therapy for von Gierke's Disease: A Case Report. Cureus. 2017 Aug 8;9(8):e1548. doi: 10.7759/cureus.1548. PMID: 29018645; PMCID: PMC5630462.

Word Meaning

Hypoglycemia: This is the condition where blood glucose level falls below 40 mg/dl. If not treated timely, this may prove fatal as neurons does not get adequate glucose and they die.

CASE 10

Cori Disease/Forbes Disease/ Limit Dextrinosis (Glycogen Storage Disorder Type III)

A 3-year-old female child born to consanguineous marriage is admitted with progressive abdominal distension for past 8 months. Mother reported that baby has difficulty in walking and gets easily tired while playing. Mother also gives the history that at times baby gets very restless and sweat in early morning hours. After giving juice or milk, she feels comfortable and sleep.

On examination, child was found to have motor retardation and was small for age. Liver was found to be enlarged (10 cm below the costal margin) with firm consistency.

Following are the results of various laboratory investigations:

AST: 345 IU/L (normal = 15–45 IU/L)

ALT: 234 IU/L (normal = 15–45 IU/L)

ALP: 221 IU/L (normal = 98–279 IU/L)

Total bilirubin: 1.8 mg/dl (normal = 0–1.0 mg/dl)

Fasting blood glucose: 56 mg/dl (normal = 80–110 mg/dl)

Liver histology revealed no fibrosis or inflammatory cell infiltration, but glycogen laden vacuoles were seen.

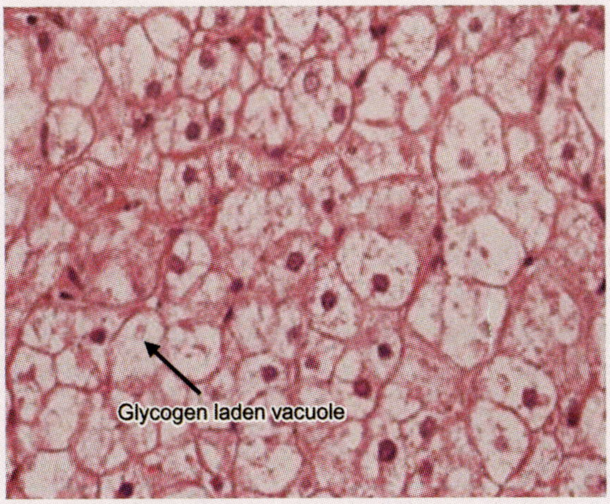

Glycogen laden vacuoles in Cori disease

Attending pediatrician advised **glucagon challenge test** which mother refused. Baby was diagnosed GSD type III based on liver histology and enzyme analysis.

QUESTIONS

Q.1. What is the causative enzyme deficiency in this disease?

Q.2. Why there is muscle involvement (myopathy)?

Q.3. What type of glycogen is expected in liver cells in biopsy?

Q.4. What is the glucagon challenge test? What is the rationale behind glucagon challenge test?

Q.5. What is the dietary advice given to mother?

Explanations

Ans.1. Enzyme deficient in **type III GSD is alpha-1,6-glucosidase or debranching enzyme** which is responsible for release of last exposed glucose residue from a glycogen chain. In deficiency of this enzyme, glycogen polymer with small branches which structurally resemble limit dextrin tend to accumulate in liver and muscle, hence this disease is called limit dextrinosis.

Ans.2. Muscle is involved due to incomplete breakdown of glycogen in muscle. Glucogenolysis in skeletal muscle provides glucose 6-phosphate which enters glycolytic pathway and generates ATP to sustain exercise. Any metabolic error in muscle glycogen breakdown results in exercise intolerance.

Ans.3. **Limit dextrin** type of molecule tends to accumulate in liver cell. This is due to normal action of **phosphorylase enzyme** and lack of **debranching activity.** This results in formation of limit dextrin which tends to accumulate over time.

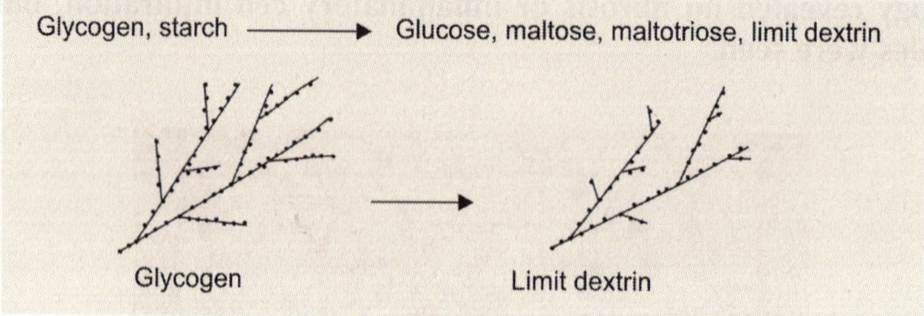

Fig. 1: Formation of limit dextrin after glycogen breakdown

Ans.4. As the symptoms of type I and type III GSD are overlapping (in GSD type III symptoms are similar to GSD I but less severe), the differentiation between type I and type III GSD is made by the **intravenous glucagon test** in the immediate postprandial period when there is usually a rise in blood sugar levels in type III but not in type I patients.

Glucagon Challenge Test

There is rise in blood sugar immediately after giving glucagon in postprandial state but there occurs no increase in blood sugar in giving glucagon in fasting state. This is due to the fact that glucagon increases the action of phosphorylase enzyme via phosphorylation, but this phosphorylase has action only on alpha 1,4-glycosidic linkages to release the glucose. Well-formed glycogen with intact alpha 1,4-glycosidic linkages are available in postprandial state only, hence infusion of glucagon in well fed state tends to increase glucose in the blood.

In fasting state, the glycogen is already broken by phosphorylase enzyme and so the infusion of glucagon has no effect in increasing the blood glucose.

In other words, in fasting state, the glycogen is already degraded and is present in limit dextrin form and glucagon administration though activates phosphorylase enzyme, it does not get the terminal glucose which can be released.

Ans.5. **Dietary advice is given in:**

- High protein diet.
- Uncooked starch to maintain blood glucose level.
- Gastric drip at night, if hypoglycemia is frequently reported.
- Frequent small meal with snacks in between.

Bibliography

1. Alvear Ciro C, Barboza Miriam, Rodríguez Zeudy K. Glycogen storage disease: report of two cases in the city of CartagenaEnfermedades de depósito de glucógeno: informe de dos casosen la ciudad de Cartagena. Colomb. Med. [Internet]. 2010 Mar [cited 2020 Dec 05] ; 41(1): 76–81.
2. Zimakas PJ, Rodd CJ. Glycogen storage disease type III in Inuit children. CMAJ. 2005 Feb 1;172(3):355–8. doi: 10.1503/cmaj.1031589. PMID: 15684118; PMCID: PMC545759.

Word Meaning

Consanguineous marriage: Marriage between cousins is known as consanguineous marriage. It is a type of interfamilial union defined as marriage between individuals who are second cousins or closer.

Parental consanguinity is associated with increased risk of autosomal recessive disorders and congenital abnormalities in the offsprings.

Limit Dextrin: Limit dextrin is an oligosaccharide which has got 8 to 10 residues of glucose in a slightly branched fashion.

It is the digestive end product of starch and glycogen in the human intestine.

CASE

11

McArdle Disease (Glycogen Storage Disorder Type V)

A 12-year-old boy presents to the OPD with complain of muscle cramps and stiffness in lower limbs soon after beginning of exercise since four years which has worsened in past 6 months. He easily gets tired while playing with friends. On careful history taking, he reveals that he feels better after 10 minutes of continuous exercise.

On examination, low lactate level in the blood was found after exercise. Muscle biopsy shows high content of glycogen (4% *vs* 0.7%).

Following are the results of various laboratory investigations:

- Blood glucose: 100 mg/dl (normal random = 80–140 mg/dl)
- Blood lactate: 0.3 mmol/L (normal level of blood lactate is 0.5–1 mmol/L)
- Creatine phosphokinase (CPK) total: 2289 IU/L (normal = 40–300 IU/L)
- Myoglobin: 556 ng/ml (normal = 25–70 ng/ml).

QUESTIONS

Q.1. What may be the probable diagnosis in this child?

Q.2. What is the biochemical basis of the signs and symptoms? What is second wind phenomenon?

Q.3. What treatment is advised for this patient?

Q.4. Whether this patient can use blood glucose for ATP production or not?

Q.5. What is the reason of dark colour urine in this patient?

Explanations

Ans.1. This child is suffering with **McArdle disease**, a **glycogen storage disorder** type V, which is characterized by deficiency of **muscle phosphorylase enzyme.** It is an **autosomal recessive** disorder which is due to mutation of **myophosphorylase gene** (PYGM) located on **chromosome 11.** This disease is named after **Dr Brian McArdle** who described it in the year 1951. Most of the cases are reported in adolescence or late adulthood.

Ans.2.

Biochemical Basis of Signs and Symptoms

Following are biochemical explanations for underlying signs and symptoms.

a. ***Exercise intolerance:*** Myophosphorylase is the glycogenolytic enzyme in muscle which is responsible for cleaving terminal **alpha-1, 4-glycosidic linkage** in a glycogen polymer. This releases glucose 1-phosphate units which enter the glycolytic pathway after getting converted to glucose 6-phosphate with the help of enzyme **phosphoglucomutase.**

This disease is characterized by exercise intolerance which is due to lack of proper glycogenolysis in muscle which is not able to supply glucose 1-phosphate. Lack of glucose 1-phosphate results in lack of glucose 6-phosphate which thus is not available for glycolysis.

Pre-exercise consumption of carbohydrate is beneficial as the blood glucose can sustain the exercise for a while in such cases.

b. ***Low lactate in blood even after exercise:*** Lactate is not produced due to lack of glycolysis, resulting in low level of lactate in the blood after the exercise. Muscle shows abnormal high content of **glycogen (4%).** Muscle exercise results in necrosis and high level of **myoglobin.**

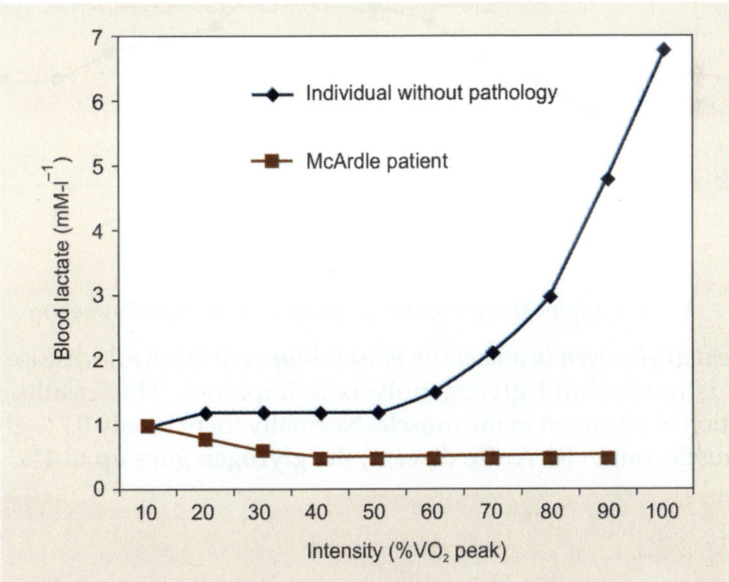

Fig. 1: Graph showing low blood lactate level even after exercise in McArdle disease

c. ***Improvement in exercise tolerance on continuing the exercise:*** To understand this, it is important to understand the second wind phenomenon. There are four important characteristics seen in McArdle disease which help in making a diagnosis:
   1. **Exercise intolerance** and low blood lactate level after exercise
   2. High total CPK level even at rest
   3. High CPK level after exercise
   4. **Second wind phenomenon**

Second Wind Phenomenon

Second wind phenomenon was first described by **Pearson** et al and it denotes considerable improvement in exercising capacity after first 10 minutes of exercise. This is explained by increased blood flow in the muscle during early phase of exercise which starts supplying fatty acid for beta oxidation and ATP production which improves the exercise tolerance. **Second wind phenomenon is pathognomonic of McArdle disease.** Figure 2 explains the second wind phenomenon graphically.

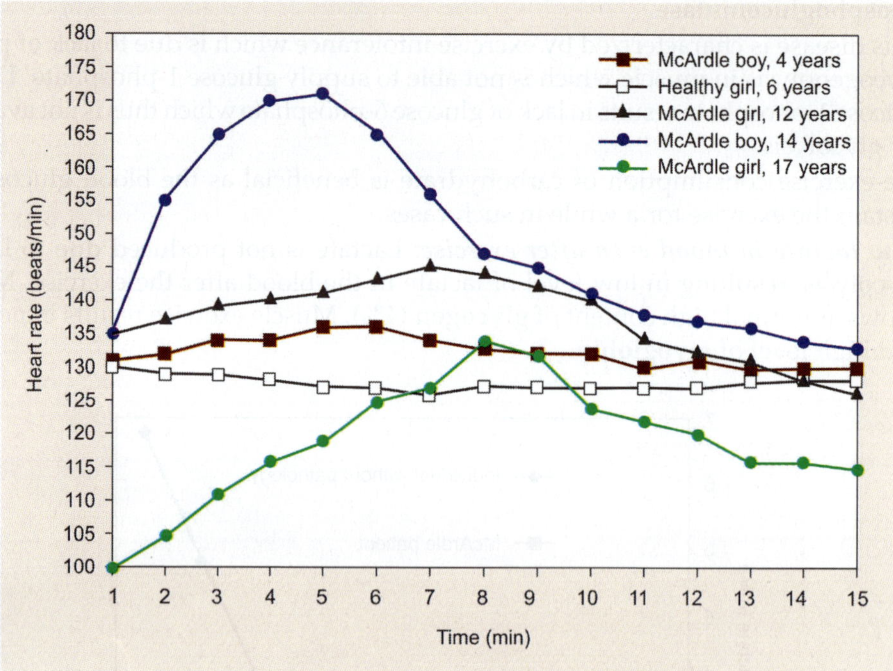

Fig. 2: Graphical representation of second wind phenomenon

d. ***High content of glycogen in muscle on muscle biopsy:*** In McArdle disease, the glycogen synthesis is normal and glycogenolysis is impaired. This results in abnormal accumulation of glycogen in the muscle. Normally there is only 0.7% glycogen in the skeletal muscle, but in **McArdle disease, the glycogen goes up to 4%.**

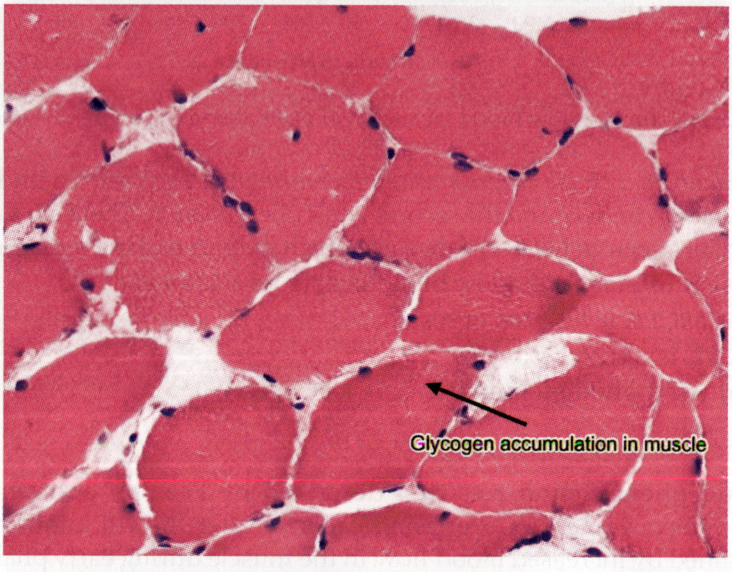

Fig. 3: High content of glycogen in the muscle in McArdle disease

It is important to note that in McArdle disease, liver glycogen metabolism is normal and there is no hypoglycemia or hepatomegaly.

Another glycogen storage disorder where glycogen phosphorylase is deficient in **Hers' disease (type VI glycogen storage disorder)**. In this disease, glycogen phosphorylase is deficient in hepatic cells alone **(Hers' Hepatic)** and muscle glycogen metabolism is normal.

In Hers' disease, presenting symptom and sign is hypoglycemia with hepatomegaly. Muscle is spared and there is no exercise intolerance.

Ans.3. The treatment for this disease is targeted at maintaining the blood glucose level throughout the day and consumption of **20 g of glucose or fructose five minutes before the start of exercise.**

Ans.4. Skeletal muscle of this patient can very well uptake blood glucose and use it in the process of glycolysis as to produce ATP.

Ans.5. Dark color urine is due to passage of myoglobin in the urine. Excessive secretion of myoglobin for a long duration may result in renal failure.

Fig. 4: Dark color urine in a case of myoglobinuria

Cycle Test

It is a physiological test which can be used to diagnose GSD V. This test record the heart rate response seen in second wind phenomenon. Here heart test is found to be notably increasing during first 10 minutes of exercise in addition to muscle cramps and pain.

Word Meaning

Rhabdomyolysis: Muscle breakdown.

Pathognomonic sign and symptoms which is specifically indicative of a disease or condition.

Bibliography

1. Leite A, Oliveira N, Rocha M. McArdle disease: a case report and review. Int Med Case Rep J. 2012 Jan 20;5:1–4. doi: 10.2147/IMCRJ.S28664. PMID: 23754915; PMCID: PMC3658246.
2. Tüzün A, Erdil A, Baðci S, Bolu E, Ongürü O, Kurt I, Yeþilova Z, Daðalp K. McArdle's disease: case report and review of the literature. Turk J Gastroenterol. 2002 Mar;13(1):56–9. PMID: 16378276.

CASE

12

Pompe Disease (Glycogen Storage Disorder Type II)

A 4-month-old child comes with history of rapid heart beat and restlessness. He was hypotonic and had hepatosplenomegaly.

He was diagnosed to have cardiac failure and was put on diuretics and **ACE inhibitor***. Chest X-ray showed **left ventricular hypertrophy***. The assay for **alpha-glucosidase** was done on dried blood spot.

The activity of the alpha-glucosidase in the patient was 0.108 μmol/L/h (normal range: 0.75–7.23 μmol/L/h)

Diagnosis of Pompe disease is made.

QUESTIONS

Q.1. Comment on the disease patient is suffering from. What enzyme is deficient in Pompe disease?

Q.2. How many clinical variants are known for Pompe disease?

Q.3. Which organ gets affected in this disorder?

Q.4. What is the treatment advised?

Explanations

Ans.1. Pompe disease is due to deficiency of lysosomal enzyme **"acid maltase (also known as acid alpha-1,4-glucosidase).** This enzyme plays important role in degradation of old and incompletely formed glycogen in the lysosome. Pompe disease is the only GSD which is included under **lysosomal storage disorder.**

This results in accumulation of glycogen granules in lysosome in virtually every tissue.

Major organ affected is heart where hypertrophic and **dilated cardiomyopathy** is observed.

Ans.2. Variants of Pompe disease are:

- Infantile
- Juvenile
- Adult

Prognosis is poor in infantile onset Pompe disease and death ensues within one year of life because of heart failure.

In adult onset Pompe disease, muscle hypotonia is seen and death is due to **respiratory insufficiency rather than heart failure.**

Ans.3. Key organ affected is **heart and respiratory muscles.** This disease is rapidly progressive and death ensues within first year of life.

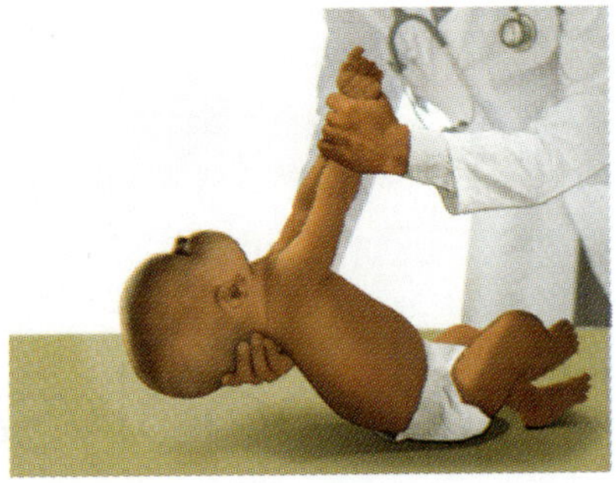

Fig. 1: Hypotonia observed in Pompe disease

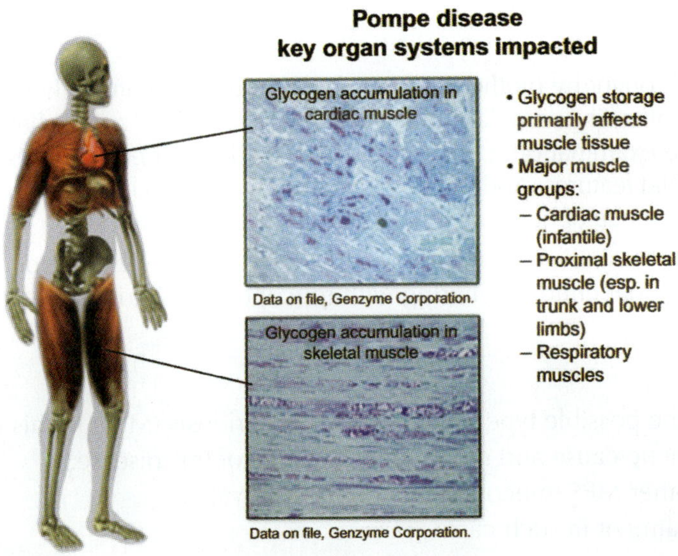

Ans.4. **Enzyme replacement therapy (ERT)** is seen to prolong life. ERT can be introduced with standard dose of 20 mg/kg every 2 weeks. Individual response to such therapy varies and largely depends on age of presentation. Literatures have shown considerable improvement in cardiac involvement after such treatment.

Bibliography

1. https://www.jpma.org.pk/PdfDownload/2939
2. Jegadeeswari A, Amuthan V, Janarthanan RA, Murugan S, Balasubramanian S. Two cases of Pompe's disease: case report and review of literature. Indian Heart J. 2012 Mar-Apr;64(2):214-6. doi: 10.1016/S0019-4832(12)60067-4. Epub 2012 Apr 28. PMID: 22572506; PMCID: PMC3861064.
3. Liu Y, Yang Y, Wang B, Wu L, Liang H, Kan Q, Cao Z, Zhao Y, Zhou X. Infantile Pompe disease: A case report and review of the Chinese literature. Exp Ther Med. 2016 Jan;11(1):235-238. doi: 10.3892/etm.2015.2862. Epub 2015 Nov 12. PMID: 26889246; PMCID: PMC4726885.

Word Meaning

ACE inhibitor: Angiotensin-converting enzyme inhibitor.

Left ventricular hypertrophy: Thickened and enlarged left ventricle.

CASE

13

Hurler Syndrome (Mucopolysaccharidosis Type I)

A 5-year-old boy is brought by mother with complain of difficulty in walking and joint stiffness for past 6 months. She complains that abdomen is protruded and he has difficulty in seeing objects. On examination, child had short neck, low set ears, depressed nasal bridge and coarse facial features. Abdominal examination revealed hepatosplenomegaly and urine showed presence of heparan sulfate and dermatan sulfate.

Enzyme assay in fibroblast reveals deficiency of L-iduronidase. Attending pediatrician came to a conclusion that child is suffering with mucopolysaccharidosis (MPS).

QUESTIONS

Q.1. What may be the possible type of mucopolysaccharidosis (MPS) in this case?
Q.2. Explain the genetic cause and mode of inheritance of this disease.
Q.3. What are the other MPS (mucopolysaccharidosis)?
Q.4. What is the treatment in such case?

Explanations

Ans.1. This child is suffering with **MPS I (Hurler syndrome)** as diagnosed by deficiency of **L-iduronidase** in the **fibroblast.** Due to deficiency of this enzyme, the **glycosaminoglycans**, which are accumulated in lysosomes, are **heparan sulfate** and **dermatan sulfate** which come out in the urine.

MPS is heterogenous group of disorder where various mucopolysaccharides are accumulated in lysosome in tissues like heart, respiratory system, bones, joints and central nervous system due to deficiency of one or other enzyme involved in degradation of these compounds in lysosome. Such abnormal accumulation of these compounds results in relevant clinical manifestation. **Corneal clouding** and **macroglossia** with **spaced dentition** are **characteristic findings** (Figs 1 and 2).

Definitive diagnosis is usually possible through enzymatic assays of the defective enzyme in cultured fibroblasts or leukocytes.

Fig. 1: Corneal clouding in Hurler's disease

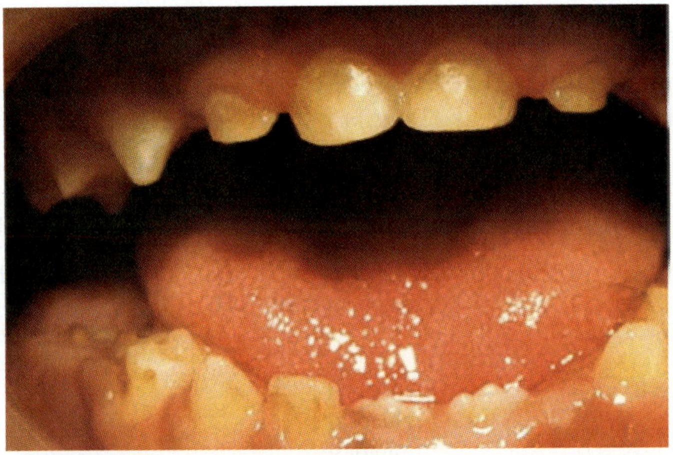

Fig. 2: Macroglossia and spaced dentition in a child with Hurler syndrome

Ans.2. MPS I is **autosomal recessive** disorder which is due to **mutation of gene IDUA** located at short arm of **chromosome 4 (4p16.3).** This results in deficiency of enzyme **alpha-L-iduronidase.**

It was first described by **Gertrude Hurler in 1919** and later the milder form of the disease was described by ophthalmologist **Dr Scheie in 1962.** Depending upon the severity of the disorder, type I MPS has three subtypes:

- **MPS I-H: Most severe**
- **MPS I-H/S: Intermediate form**
- **MPS I-S: Mild form**

Ans.3. There are total 7 types of MPS. They are described in Table 1.

Table 1: Various mucopolysaccharidoses, deficient enzymes and clinical features

Type	*Name of syndrome*	*Incidence*	*Deficient enzyme*	*Accumulated products*	*Main symptoms*
I	Hurler (I-H) (most severe form)	1/100.000	Alpha-L-iduronidase	HS, DS	Mental retardation; corneal clouding, photophobia; organomegaly; heart disease; death in early childhood
	Scheie (I-S) (mildest form)				Corneal clouding; stiff joints; normal intelligence and life-span
	Hurler-Scheie (I-H/S) (intermediate form)				Intermediate phenotype, between MPS I-H and MPS I-S

Contd..

Table 1: Various mucopolysaccharidoses, deficient enzymes and clinical features *(Contd.)*

Type	*Name of syndrome*	*Incidence*	*Deficient enzyme*	*Accumulated products*	*Main symptoms*
II	Hunter	1/250.000	Iduronate sulfatase	HS	Cornea clear, organomegaly; short stature; death before 15 years
III	Sanfilippo A	1/150.000	Heparan sulphamidase	DS HS	Mild somatic manifestations; hyperactivity; profound intellectual deterioration
	Sanfilippo B		N-acetyl-glucosaminidase		
	Sanfilippo C		Acetyl-CoA: Alpha-glucosaminide-acetyl transferase		
	Sanfilippo D		N-acetyl-glucosamine 6-sulfatase		
IV	Morquio A	1/75.000	Galactose-6-sulfate sulfatase	KS	Short stature; motor dysfunction
	Morquio B		Beta-galactosidase	CH KS	
V	The MPS I-S was initially described as type V. This is no more being used as separate entity.		Same as type I-S		
VI	Maroteaux-Lamy	<1/250.000	N-acetylgalactosamine-4-sulfatase	DS	Short stature; motor dysfunction; kyphosis; heart defects; death in early childhood
VII	Sly	<1/250.000	Beta-glucuronidase	HS, DS,CH	Hepatomegaly; dysostosis multiplex; short stature; corneal clouding; develop mental delay

Ans.4. Treatment is **multidisciplinary**. Following plan is adopted:

- Hematopoietic stem cell transplantation (HSCT)
- Enzyme replacement therapy (ERT)
- Supportive therapy

- Physical therapy
- Occupational therapy
- Speech therapy
- Respiratory support

Bibliography

1. Sharma S, Sabharwal JR, Datta P, Sood S. Clinical manifestation of Hurler syndrome in a 7-year-old child. Contemp Clin Dent. 2012 Jan;3(1):86–9. doi: 10.4103/0976-237X.94554. PMID: 22557905; PMCID: PMC3341767.
2. Tatapudi R, Gunashekhar M, Raju PS. Mucopolysaccharidosis type I Hurler-Scheie syndrome: A rare case report. Contemp Clin Dent. 2011 Jan;2(1):66–8. doi: 10.4103/0976-237X.79287. PMID: 22114460; PMCID: PMC3220181.

CASE

14

Hunter Syndrome (Mucopolysaccharidosis Type II)

A 3-year-old male child is brought by mother with history of pain in lower limb and difficulty in walking for past 4 months. History of delayed milestone present.

On examination, child had coarse facial features and low IQ. Liver was found to be enlarged, 4 cm below the costal margin. Cornea was clear and child had stiffness of joints in hand and lower limb.

Urinary examination revealed presence of heparan sulfate and dermatan sulfate and fibroblast showed deficiency of enzyme iduronate-2-sulfatase.

Diagnosis of Hunter syndrome was made.

QUESTIONS

Q.1. Comment on the diagnosis.

Q.2. What is the inheritance of Hunter's disease?

Q.3. What is the treatment plan in such cases?

Explanations

Ans.1. **Hunter disease is MPS II** which is caused by deficiency of enzyme **iduronate-2-sulfatase** resulting in accumulation of **MPS (HS, DS) in various tissues.** This disease is named after **Charles Hunter,** who first described the condition as an **X-linked recessive disease** in the year 1917.

Characteristic finding in this disease is mostly coarse **facial features, macroglossia, short stature, joint contractures, visceromegaly, intellectual disability and dysostosis multiplex*.** In addition, **sleep apnea, cardiomyopathy, cardiac valve dysplasia,** and **dental and dermatological manifestations** may be present.

Presentation is in early childhood as developmental delay and organomegaly, though the milder cases may present in adulthood as joint contracture and hoarseness of voice.

Ans.2. Hunter disease is a rare disorder which is inherited in **X-linked recessive fashion**. Gene is located at **Xq28.** Male child is affected predominantly though reports of this disease being present in female child is also present. Overall incidence of this disease is 1.3/1,00,000 male live births.

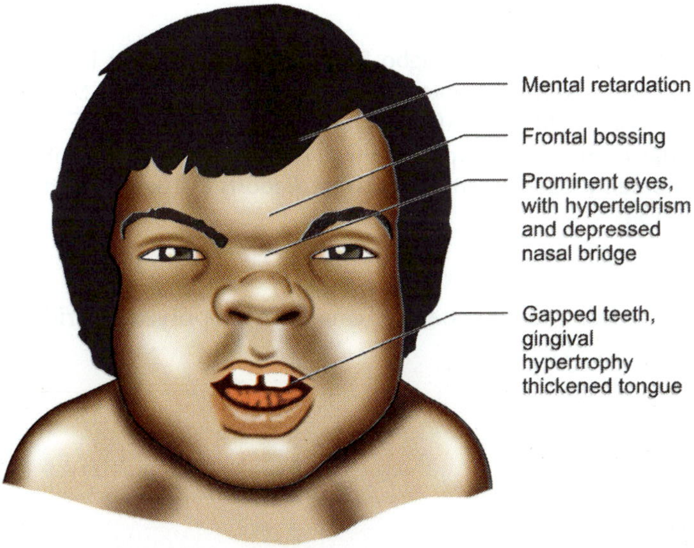

Fig. 1: Facial features in Hunter syndrome

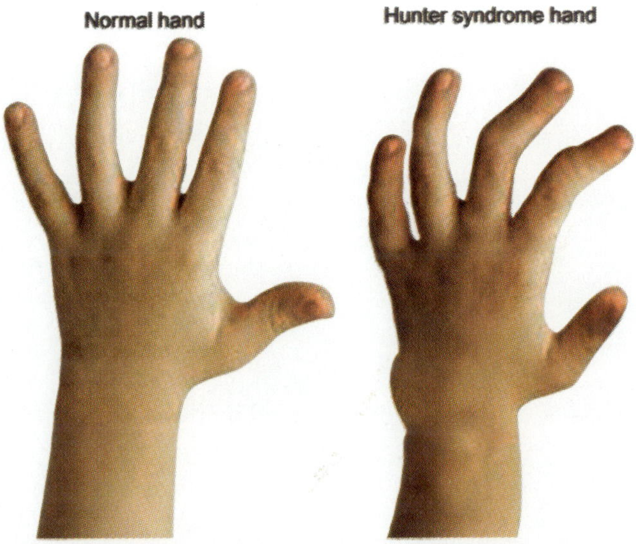

Fig. 2: Joint contracture and claw hand in Hunter syndrome

In fact, Hunter disease is the only MPS which has X-linked recessive inheritance. All other MPS described in Table 1 (Case 13) has autosomal recessive inheritance.

Ans.3. Management is multidisciplinary and a holistic approach to the patient is required, especially for patients with severe neurological involvement. Early diagnosis and initiating early treatment, particularly enzyme replacement therapy (ERT), may reduce possible disease complications and can greatly improve quality of life.

Bibliography

1. Chinawa J, Adimora G, Obu H, Tagbo B, Ujunwa F, Onubogu I. Clinical Presentation of Mucopolysaccharidosis type II (Hunter's syndrome). Ann Med Health Sci Res. 2012 Jan;2(1):87–90. doi: 10.4103/2141-9248.96946. PMID: 23209998; PMCID: PMC3507130.
2. Shah GS, Mahal T, Sharma S. Atypical clinical presentation of mucopolysaccharidosis type II (Hunter syndrome): a case report. *J Med Case Reports* 2010;4:154. https://doi.org/10.1186/1752-1947-4-154.

Word Meaning

Dysostosis multiplex: Skeletal changes due to mucopolysaccharidosis.

CASE 15

Zellweger Syndrome

A 20-day-old newborn of normal health parents with seizure, low muscle tone and refusal to feed. Child was apparently normal at birth and these symptoms slowly evolved over couple of days. He has two other siblings who are normal.

On examination, baby's face was seen to be flattened with **broad nasal bridge**, **upslanting palpebral fissure and cataract** in both the eyes. USG showed **multiple cysts** in both the kidneys.

Blood showed raised *level of very long chain fatty acid, low level of glycerol-ether lipid* ***(plasmalogen)*** *and increased concentration of* ***cholestenoic acid (precursor of bile acid).***

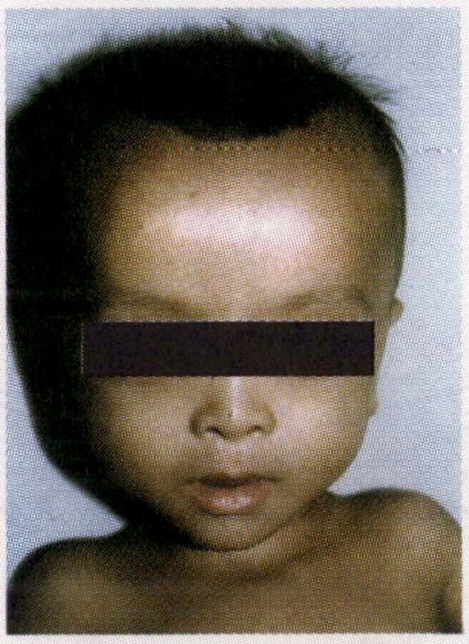

Facial appearance of a child suffering with Zellweger syndrome

Based on clinical presentation and biochemical profile, diagnosis of **Zellweger syndrome** is made.

QUESTIONS

Q.1. What is the biochemical defect in Zellweger syndrome?

Q.2. How do you explain the biochemical blood profile of this patient?

Q.3. What all organs may be involved in this case?

Q.4. What is the mode of inheritance? In other word, what is the probability of other sibling getting affected with this disorder?

Q.5. What is the current treatment strategy in this case?

Q.6. What are other diseases which are included in peroxisomal biogenesis disorder (PBD)?

Explanations

Ans.1. **Zellweger syndrome** is the most severe presentation of a group of disorder known as **"peroxisomal biogenesis disorder (PBD)".** PBD comprises more than 25 genetically related disorders where function of peroxisome is either partially or fully compromized. Most common gene involved in Zellweger syndrome is ***PEX1 gene.***

Ans.2. In Zellweger syndrome, lack of peroxisome is seen which results in lack of all the functions for which peroxisome is required, e.g. VLCFA oxidation, bile acid synthesis, branched-chain fatty acid oxidation, and synthesis of ether lipid (plasmalogen).

Ans.3. Zellweger syndrome is also called **cerebrohepatorenal syndrome** because of involvement of these organs. Eyes, liver, kidney, brain involvement are commonly seen.

Ans.4. It is an **autosomal recessive disorder.**

So, parents of affected child are carrier and 1 in 4 chance baby will develop the disease. 2 of 4 child will be carrier and 1 of 4 child will be normal.

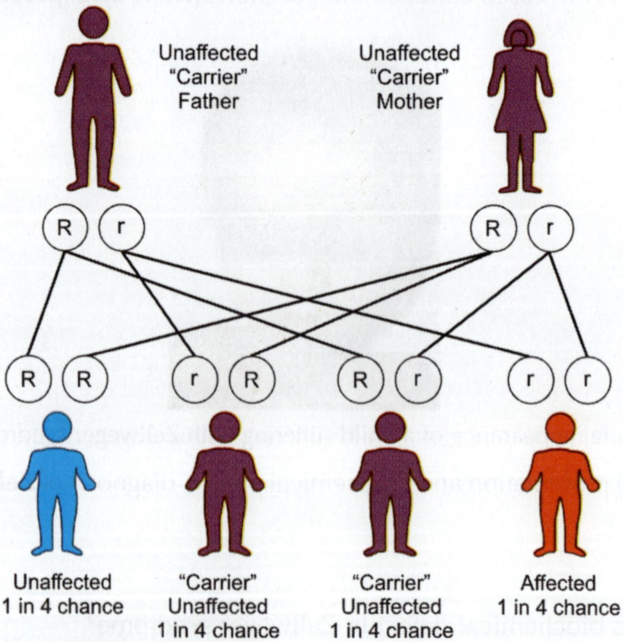

Fig. 1: Autosomal recessive inheritance

Ans.5. Currently no definitive treatment is available for this disorder. Symptomatic treatment is advised. Prognosis is very poor and in most of the cases, baby dies within one year of life.

Ans.6. Zellweger syndrome is the most severe presentation of a group of disorder known as "**peroxisomal biogenesis disorder (PBD)**".

Intermediate form of PBD is **X-linked adrenoleukodystrophy (XALD)** where defect lies in VLCFA transporter protein **(adrenoleukodystrophy/ALD protein)** in the peroxisomal membrane. This affects young boys and VLCFA accumulates as it is not oxidized, remaining function of peroxisome as described earlier remains intact.

Refsum's disease (described in next case) is mildest form of PBD where the enzyme **'phytanoyl-CoA hydroxylase' is defective** resulting in impaired oxidation of branched chain fatty acid **(phytanic acid)**. This disease presents with severe neurological dysfunction along with blindness and deafness.

Bibliography

1. Galvez-Ruiz A, Galindo-Ferreiro A, Alkatan H. A clinical case of Zellweger syndrome in a patient with a previous history of ocular medulloepithelioma. Saudi J Ophthalmol. 2018 Jul-Sep;32(3):241–45. doi: 10.1016/j.sjopt.2017.09.004. Epub 2017 Sep 23. PMID: 30224891; PMCID: PMC6137698.
2. Lee HF, Mak SC, Wu FW, Chi CS, Huang SC. Zellweger syndrome: report of one case. Acta Paediatr Taiwan. 2001 Jan-Feb;42(1):53–6. PMID: 11270189.

CASE

16

Refsum Disease

A 2½-year-old girl child is brought to hospital by parents with complain of difficulty in swallowing food, impaired vision specially at night and hard of hearing. She was born to healthy parents. Pregnancy and delivery was uneventful. She was apparently normal at birth but these symptoms worsen over past six months.

On examination, she was seen to have bilateral nystagmus. On examination of both the feet, 4th phalange was short. Skin was dry. Electroretinogram showed retinal pigmentation in both the fundi.

Following are the results of various laboratory investigations:

Elevated VLCFA: Cerotic acid (C26:0) was 1.33 μg/ml (normal = 0.22 μg/ml)

Elevated phytanic acid: 5.66 μg/ml (normal = 0.12 μg/ml)

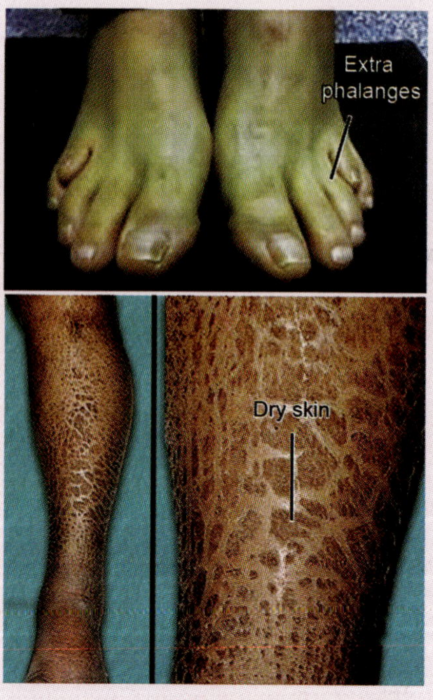

The diagnosis of infantile Refsum's disease is made.

QUESTIONS

Q.1. Comment on the diagnosis.

Q.2. What is the source of phytanic acid?

Q.3. What is the current management strategy?

Explanations

Ans.1. **Infantile Refsum's disease** is an inborn error of metabolism which is due to defect in alpha oxidation of phytanic acid. Refsum's disease was first described by Refsum in 1940s. It is an autosomal recessive condition.

Phytanic acid is a branched-chain fatty acid, oxidation of this fatty acid requires alpha oxidation followed by beta oxidation in sequential manner. Due to deficiency of this enzyme, alpha oxidation does not take place and beta oxidation cannot proceed as usual resulting in accumulation of phytanic acid in various organs.

In **Refsum's disease**, it is not the beta oxidation which is at fault, rather it is alpha oxidation which is defective as one of the enzymes of **alpha oxidation** known as **phytanic acid alpha oxidase/phytanic acid alpha hydroxylase** is deficient.

Clinically, it is manifested by mental and growth retardation, facial dysmorphism, hepatomegaly, severe **sensorineural deafness**, atypical **retinitis pigmentosa**, peripheral neuropathy, cerebellar ataxia, and high concentrations of protein in CSF.

Ans.2. **Phytanic acid** is **exogenous, milk and dairy products** are primary sources of phytanic acid. It is also found abundantly in plants but not available for absorption from these sources as chlorophyll make insoluble complex which prevents its absorption from gastrointestinal lumen.

Ans.3. Currently symptomatic management is done. There is no cure to this disease till date.

Bibliography

1. Marano R, Soliveri P, Garavaglia B, *et al.* A case of Refsum disease with atypical clinical picture in family members. *Ital J Neuro Sci* 1989;10:451–54. https://doi.org/10.1007/BF02334952.

CASE

17

Respiratory Distress Syndrome (RDS)

Preterm neonate born to unbooked mother* shows increased respiratory rate, chest wall retraction, cyanosis*, pallor*, his condition was progressively worsening and he was put on oxygen.

X-ray chest was showing ground glass appearance. Attending pediatrician suspected **respiratory distress syndrome** (RDS).

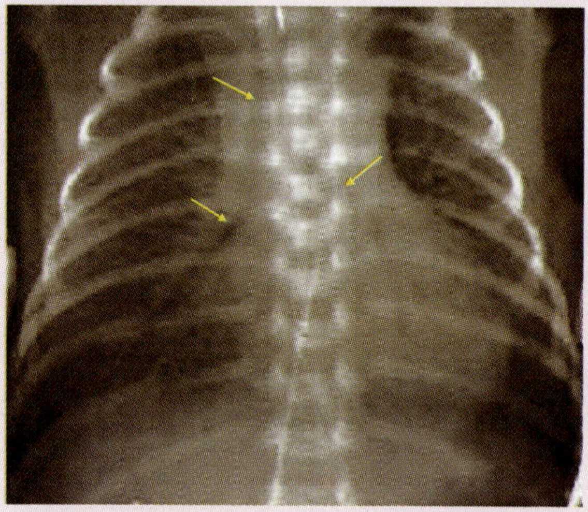

X-ray chest showing ground glass appearance

QUESTIONS

Q.1. What is the cause of respiratory distress syndrome (RDS) in premature babies?

Q.2. What is the major component of lung surfactant?

Q.3. How to assess the lung maturity antenatally?

Q.4. What precaution to be taken antenatally to prevent RDS?

Explanations

Ans.1. RDS (respiratory distress syndrome) is an important cause of neonatal morbidity and mortality in both developing and developed countries. RDS affects premature babies, because of immaturity of their lungs which do not produce sufficient lung surfactant.

Lung surfactant is an important substance which is layered on internal wall of alveoli and reduces surface tension in the alveoli in such a manner that they are not collapsed and shape of alveoli is maintained for proper gaseous exchange.

Ans.2. Major component of lung surfactant is a **phospholipid (dipalmitoyl lecithin)**, which prevents the collapse of lung alveoli during expiration. Its production increases as the gestational age advances.

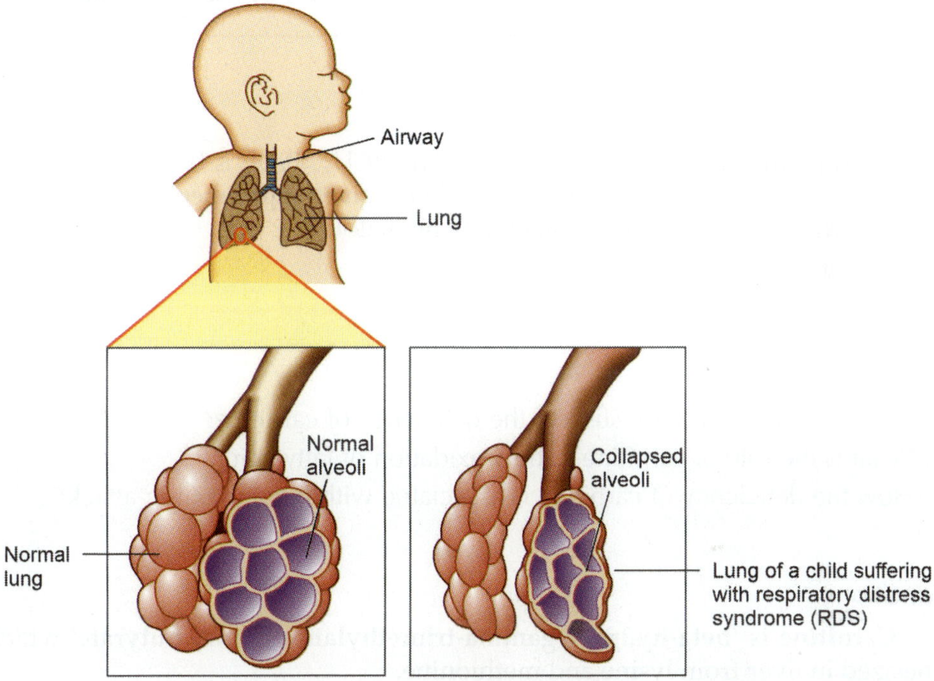

Fig. 1: Collapsed lung in RDS

Ans.3. Lung maturity in fetus is assessed by **lecithin/sphingomyelin ratio (L/S ratio)** in amniotic fluid. **L/S ratio of "2" denotes lung maturity** and is achieved at gestational age of 34 weeks. Risk of developing **RDS** is >75%, if L/S ratio is <1.5.

Ans.4. Glucocorticoid injection given antenatally to mother to enhance fetal lung maturity.

Bibliography

1. Laird P, Ruppert SD. Acute respiratory distress syndrome—a case study. Crit Care Nurs Q. 2011 Apr-Jun;34(2):165–74. doi: 10.1097/CNQ.0b013e3182129de2. PMID: 21407013.

Word Meaning

Unbooked mother: Pregnant woman who has not attended any antenatal clinic session.

Cyanosis: Bluish skin due to poor oxygenation.

Pallor: An unhealthy pale skin.

CASE

18

Carnitine Deficiency in Premature Baby

A 6-day-old premature newborn is admitted in NICU*. Baby is born at 7 months of gestational age and is in poor condition. He is being given **supplementation of carnitine** to prevent hypoglycemic attack by attending neonatologist.

QUESTIONS

Q.1. What is carnitine?

Q.2. What all conditions you suspect the deficiency of carnitine?

Q.3. What is the role of carnitine in beta oxidation of fatty acid?

Q.4. How the deficiency of carnitine is associated with hypoglycemic attack?

Explanations

Ans.1. **Carnitine** is **'beta-hydroxy-gamma-trimethylammonium butyrate'** which is synthesized in liver from lysine and methionine.

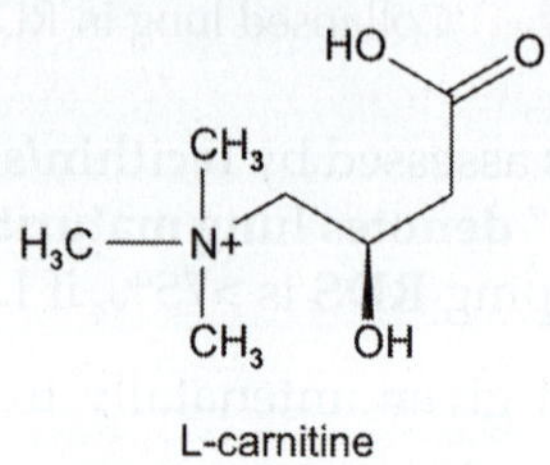

Structure of L-carnitine

Ans.2. **Conditions which may be associated with carnitine deficiency** are as follows:

a. Premature birth

b. Dialysis

c. End stage liver failure

In preterm infant, carnitine is not adequately synthesized due to immature liver and whatever small amount is synthesized is lost via kidney due to immature basement membrane.

This results in severe deficiency of carnitine in such preterm infants.

Ans.3. Carnitine is required for fatty acid transport across mitochondrial membrane. The whole process is known as carnitine shuttle.

Figure 1 is explaining the carntine shuttle. One can observe the role of carnitine in converting acyl-CoA to acylcarnitine with the help of CPT-1* at outer mitochondrial membrane.

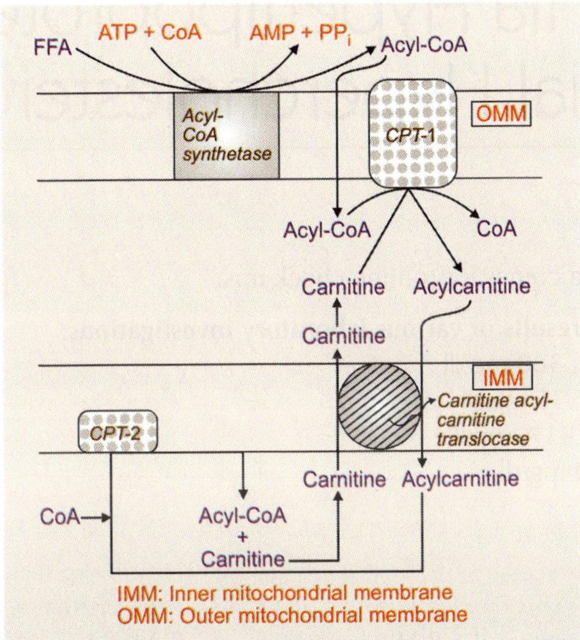

Fig. 1: Fatty acid transportation across mitochondrial membrane

Ans.4. Carnitine is important for facilitating fatty acid oxidation via helping their transportation across mitochondrial membrane. Fatty acid oxidation is important prerequisite for gluconeogenesis. Hence, deficiency of carnitine in preterm babies, gluconeogenesis is affected and such babies develop repeated **hypoglycemic episodes.**

Supplementation of carnitine in preterm babies is important to avoid hypoglycemic complications.

Bibliography

1. Rajakumari A. Acute respiratory distress syndrome: A case presentation. Indian J Cont Nsg Edn 2017;18:32–8.
2. Sweet LR, Keech C, Klein NP, Marshall HS, Tagbo BN, Quine D, Kaur P, Tikhonov I, Nisar MI, Kochhar S, Muñoz FM. Brighton Collaboration Respiratory Distress in the Neonate Working Group. Respiratory distress in the neonate: Case definition and guidelines for data collection, analysis, and presentation of maternal immunization safety data. Vaccine. 2017 Dec 4;35(48 Pt A):6506–6517. doi: 10.1016/j.vaccine.2017.01.046. PMID: 29150056; PMCID: PMC5710987.

Word Meaning

NICU: Neonatal intensive care unit.

*CPT-1: Carnitine Palmitoyl Transferase I.

CASE

19 Type IIa Hyperlipoproteinemia (Familial Hypercholesterolemia)

A 37-year-old male comes for routine check ups.

Following are the results of various laboratory investigations:

- Total cholesterol: 380 mg/dl
- HDL cholesterol: 35 mg/dl
- LDL cholesterol: 317 mg/dl
- Triglyceride: 140 mg/dl
- VLDL: 28 mg/dl

QUESTIONS

Q.1. What all parameters should be tested in extended lipid profile?
Q.2. What are the normal values of various parameters tested under lipid profile?
Q.3. What is the probable dyslipidemia this patient must be having?
Q.4. What is the difference in type IIa and type IIb?
Q.5. How to calculate LDL value, if values of total cholesterol, HDL, VLDL are given?

Explanations

Ans.1. Following are the list of tests which are being done in under **extended lipid profile:**

- Total cholesterol
- HDL cholesterol
- LDL cholesterol
- Triglyceride
- VLDL
- **Apo B**
- **Apo A**
- **LDL/HDL ratio**
- TC/HDL ratio
- **Lipoprotein (a)**

Ans.2. Normal lipid profile of adult is enlisted in Table 1.

Table 1: Values of various lipid parameters in a healthy adult

Parameters	*Normal healthy adult*
Total cholesterol	Desirable: <200 mg/dl Borderline high: 200 to 239 mg/dl High: >240 mg/dl
HDL cholesterol	40–75 mg/dl
LDL cholesterol	Optimal: <100 mg/dl Near optimal: 100 to 129 mg/dl Borderline high: 130 to 159 mg/dl High: 160 to 189 mg/dl Very high: >190 mg/dl
Triglyceride	60–150 mg/dl
VLDL	<40 mg/dl
Free fatty acid	200–800 μmol/L
Lipoprotein (a)	0 to 30 mg/dl

Ans.3. The patient in given case is suffering with **type IIa hyperlipoproteinemia (familial hypercholesterolemia)** which is due to impaired clearance of LDL from the plasma.

This may be because of **mutation of LDL receptor or may be because of mutation of Apo B100** which fail to get recognized by the **LDL receptor**. Mechanism of recognition of LDL by LDL receptor with the help of Apo B-100 is diagrammatically represented in Fig. 1.

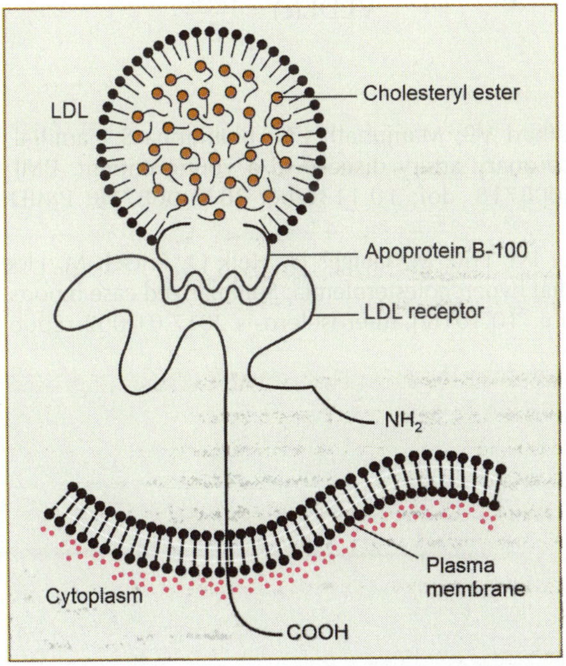

Fig. 1: Recognition of LDL particle by LDL receptor with the help of Apo B-100

Patient has high level of LDL cholesterol. The major contribution of total cholesterol is from the LDL, hence total cholesterol is also high in such patients.

Ans.4. In **type IIb hyperlipoproteinemia**; the defect of **type IIa is coexistent with excessive secretion of VLDL from liver.**

For example, the illustrative lipid profile of a patient suffering with type IIb is given below:

- Total cholesterol: 345 mg/dl
- HDL cholesterol: 42 mg/dl
- LDL cholesterol: 233 mg/dl
- Triglyceride: 350 mg/dl
- VLDL: 70 mg/dl.

This patient is suffering with **type IIb hyperlipoproteinemia (familial combined hyperlipidemia)** which is due to impaired clearance of LDL from the plasma and in addition excessive secretion of VLDL from liver.

Patient has high level of LDL cholesterol and as the major contribution of total cholesterol is from the LDL, total cholesterol is also high in such patients. Triacylglycerol level is also high due to excessive VLDL secretion from liver.

Ans.5. LDL can be calculated using Friedwald formula which is represented as follows:

LDL (c) = Total cholesterol – HDL (c) – VLDL (c)

Friedwald formula can be applied only when TAG value is $\leq$ 400 mg/dl. In case of TAG values higher than this, direct assessment of LDL to be done. VLDL (c) can be calculated from TAG value using following formula:

$$\text{VLDL(c)} = \frac{\text{TAG}}{5}.$$

Bibliography

1. Subramani KS, Kolhari VB, Manjunath CN, Bhairappa S. Familial hypercholesterolemia presenting with coronary artery disease in a young patient. BMJ Case Rep. 2013 Mar 20;2013:bcr2013008718. doi: 10.1136/bcr-2013-008718. PMID: 23519510; PMCID: PMC3618795.
2. Widhalm K, Benke IM, Fritz M, Geiger H, Helk O, Fritsch M, Hoermann G, Kostner G. Homozygous familial hypercholesterolemia: Summarized case reports. Atherosclerosis. 2017 Feb;257:86–89. doi: 10.1016/j.atherosclerosis.2017.01.002. Epub 2017 Jan 18. PMID: 28126585.

CASE

20

Acute Pancreatitis in Hypertriacylglyceridemia

A 45-year-old male presents to emergency department with complain of severe abdominal pain in epigastric region which is radiating to back for past 6 hours. He had two episodes of vomiting before coming to hospital. He is a chronic alcoholic indulged in alcohol intake for past 17 years.

Following are the results of various laboratory investigations:

- Total cholesterol: 194 mg/dl (normal = 200 mg/dl)
- HDL cholesterol: 32 mg/dl (normal = 40–75 mg/dl)
- LDL cholesterol: 102 mg/dl (normal = <100 mg/dl)
- Triglyceride: 300 mg/dl (normal = 60–150 mg/dl)
- VLDL: 60 mg/dl (normal = <40 mg/dl)
- Serum amylase: 1450 Somogyi units/L (normal = 40–140 Somogyi units/L)

QUESTIONS

Q.1. What may be the cause of abdominal pain in this patient?

Q.2. What is the possible type of dyslipidemia in this patient?

Q.3. In addition to amylase, what other enzyme assay may be useful in diagnosis of this condition? Also mention the sensitivity and specificity of these enzyme assays in diagnosis of acute pancreatitis.

Q.4. What all types of dyslipidemias are associated with pancreatitis?

Explanations

Ans.1. This patient is presenting with acute episode of pancreatitis as justified by clinical symptoms and raised amylase level.

Ans.2. Patient is suffering with **isolated hypertriacylglyceridemia [type I (familial hypertriacylglyceridemia) or type V hyperlipoproteinemia]. This is due to either lack of lipoprotein lipase enzyme or deficiency of Apo CII.**

This condition may be asymptomatic or may present with pancreatitis as in this case.

Ans.3. Another enzyme which can be assayed in acute pancreatitis is lipase. Amylase is sensitive but nonspecific for diagnosis of acute pancreatitis. Lipase is rather late to rise but is more specific for acute pancreatitis compare to amylase.

Ans.4. Hyperlipidemias associated with hypertriacylglyceridemias are associated with increased incidence of pancreatitis. They are:

- Type I
- Type IV
- Type V.

Bibliography

1. Weston N, Fernando U, Baskar V. Hypertriglyceridemia-induced pancreatitis. BMJ Case Rep. 2013 Feb 27;2013:bcr2013008722. doi: 10.1136/bcr-2013-008722. PMID: 23446049; PMCID: PMC3603802.

Type III Hyperlipoproteinemia (Remnant Removal Disease or Broad Beta Disease)

A 21-year-old medical student is send to biochemistry department for lipid profile assessment as he was complaining of **yellowish discoloration** of both of his palms for past six months. There is no history of exposure to chemicals and sclera and mucous membrane are normal.

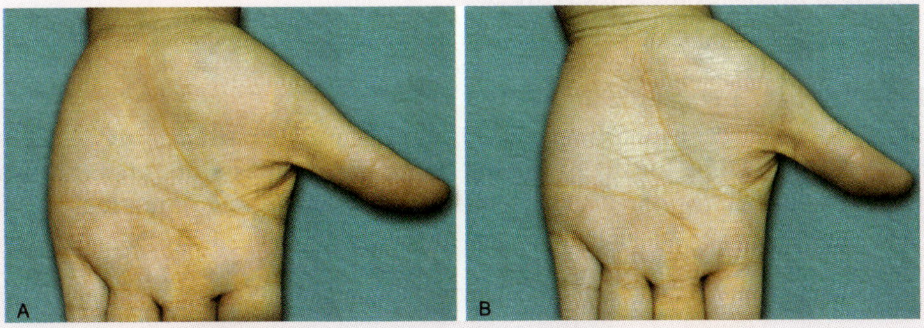

Following are the results of various laboratory investigations:
- Total cholesterol: 240 mg/dl (normal = <200 mg/dl)
- HDL cholesterol: 45 mg/dl (normal = 40–75 mg/dl)
- LDL cholesterol: 145 mg/dl (normal = <100 mg/dl)
- Triglyceride: 250 mg/dl (normal = 60–150 mg/dl)
- VLDL: 50 mg/dl (normal = <40 mg/dl)

Broad beta band was found on lipoprotein electrophoresis.

QUESTIONS

Q.1. What is the interpretation of above lipid profile?

Q.2. What is the normal pattern of lipoprotein electrophoresis. Explain your answer with suitable diagram.

Q.3. Why there is broad beta band on electrophoresis?

Q.4. How the various types of hyperlipoproteinemias are classified?

Q.5. What is abetalipoproteinemia? What is the reason behind its nomenclature?

Q.6. What will be the electrophoretic pattern in each of these types. Explain with diagram.

Explanations

Ans.1. Patient is having **type III hyperlipoproteinemia or remnant removal disease (familial dysbetalipoproteinemia).**

As the name implies, this disease is characterized by defect in removal of remnant particles of chylomicron and VLDL. For clearance of these remnant particles, presence of Apo E is required as it acts as a ligand which is identified by the receptor.

There are three types of **Apo E: E1, E2, E3.**

The one which is easily recognized by the receptor is **Apo E2**, at times known as **good Apo E**. Apo E1 and E3 are not easily recognized and are called **bad Apo E.** Individuals having remnant removal disease is either having Apo E1 or E3 hence their remnant particles are not easily cleared from circulation resulting in their accumulation in the plasma.

These remnant particles have unmetabolized triglyceride and cholesterol in them. In addition, HDL particle during its metabolism tends to exchange some of its cholesterol in these remnant particles resulting in further elevation of cholesterol. This condition is mostly asymptomatic till vascular disease develops. **Palmar or tuberous xanthomas** may be found and this is reflected in elevated triacylglycerol and cholesterol levels in lipid profile of these individuals.

Ans.2. In normal lipoprotein electrophoresis, following bands are found:

- **Alpha, pre-beta, beta and gamma**
- **HDL is localized in alpha band. VLDL and remnant particles are localized in prebeta band. LDL is localized in beta band. Chylomicron** does not move much on lipoprotein electrophoresis and tends to get localized in **gamma region.**

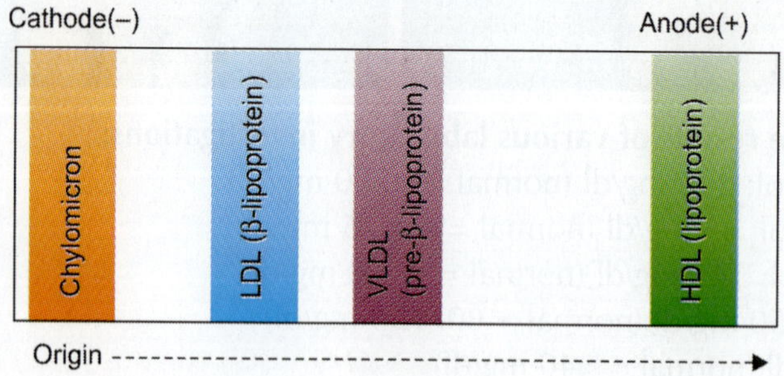

Fig. 1: Normal electrophoretic band pattern

Ans.3. In **remnant removal disease**, the excessive remnant particles which are accumulated in plasma due to their failed clearance tend to concentrate in the pre-beta band which is widened. This results in merging of beta and pre-beta band giving the name to this disease as broad beta disease.

Ans.4. Lipoproteinemias are classified as per **Fredrickson's classification (1967)**, based on electrophoretic pattern and ultracentrifugation, into following types.

Type I: Familial hypertriacylglyceridemia (familial hyperchylomicronemia/due to deficiency of **lipoprotein lipase enzyme**).

Type II: Familial hypercholesterolemia: Impaired clearance of LDL either due to LDL receptor mutation or due to mutation of **Apo B100**. (This was later subdivided into **type IIa and type IIb.** In type IIa, there is defect only in LDL clearance and in type IIb, in addition to defect in LDL clearance, there is additional VLDL overproduction.)

Type III: Remnant removal disease (familial dysbetalipoproteinemia): This is described above.

Type IV: Familial combined hyperlipidemia (There is overproduction of VLDL resulting in raised level of VLDL as well as LDL.)

Type V: Endogenous hypertriglyceridemia (This is rare and is due to deficiency of Apo CII. Apo CII is activator of lipoprotein lipase enzyme, hence, in its deficiency the lipoprotein lipase enzyme is not functional resulting in lipoprotein profile similar to type I.)

Hyperlipoproteinemia **may also be classified** based on **actual biochemical derangements** underlying them into following subtypes:

a. *Hyperlipoproteinemia caused by impaired lipolysis of triacylglycerol rich lipoprotein (type I and type V)*
b. *Hyperlipoproteinemia caused by impaired hepatic clearance of Apo B100 containing lipoprotein (type IIa and type IIb)*
c. *Hyperlipoproteinemia caused by impaired hepatic clearance of Apo E containing lipoprotein (type III/remnant removal disease)*
d. *Hyperlipoproteinemia caused by excessive hepatic secretion of triacylglycerol rich lipoprotein (type IV hyperlipoproteinemia).*

Interpretation of Lipid Profiles (Illustrative Example)

Comment on following lipid profile results of a 43-year-old labor on daily wages who otherwise is healthy and not on any medication:

- Total cholesterol: 190 mg/dl (normal = <200 mg/dl)
- HDL cholesterol: 45 mg/dl (normal = 40–75 mg/dl)
- LDL cholesterol: 75 mg/dl (normal = <100 mg/dl)
- Triglyceride: 350 mg/dl (normal = 60–150 mg/dl)
- VLDL: 70 mg/dl (normal = <40 mg/dl)

Patient is suffering with **type IV hyperlipoproteinemia.**

This is due to excessive VLDL secretion from liver due to idiopathic cause. This condition may be asymptomatic or may be associated with vascular features.

Ans.5. **Abetalipoproteinemia** is a rare autosomal recessive disorder which is characterized by the defect in incorporation of Apo B (both Apo B100 as well as Apo B48) in respective lipoprotein. This results in absolute deficiency of those lipoproteins which either contain Apo B100 or Apo B48 in the plasma. VLDL and LDL are absent as they contain Apo B100, chylomicrons are also absent as they contain Apo B48. When there is no VLDL and no CM, there is no remnant particles of these lipoproteins in the plasma. This results in lack of all lipoproteins in the plasma except HDL which is normal.

When the electrophoresis is done in plasma of this patient, it shows only alpha band which contains HDL. Remaining all other bands (pre-beta, beta, gamma) are absent. As there is no beta or pre-beta band on electrophoresis, this disease is named abetalipoproteinemia.

Ans.6. The electrophoresis patterns of various hyperlipoproteinemias are illustrated in Fig. 2.

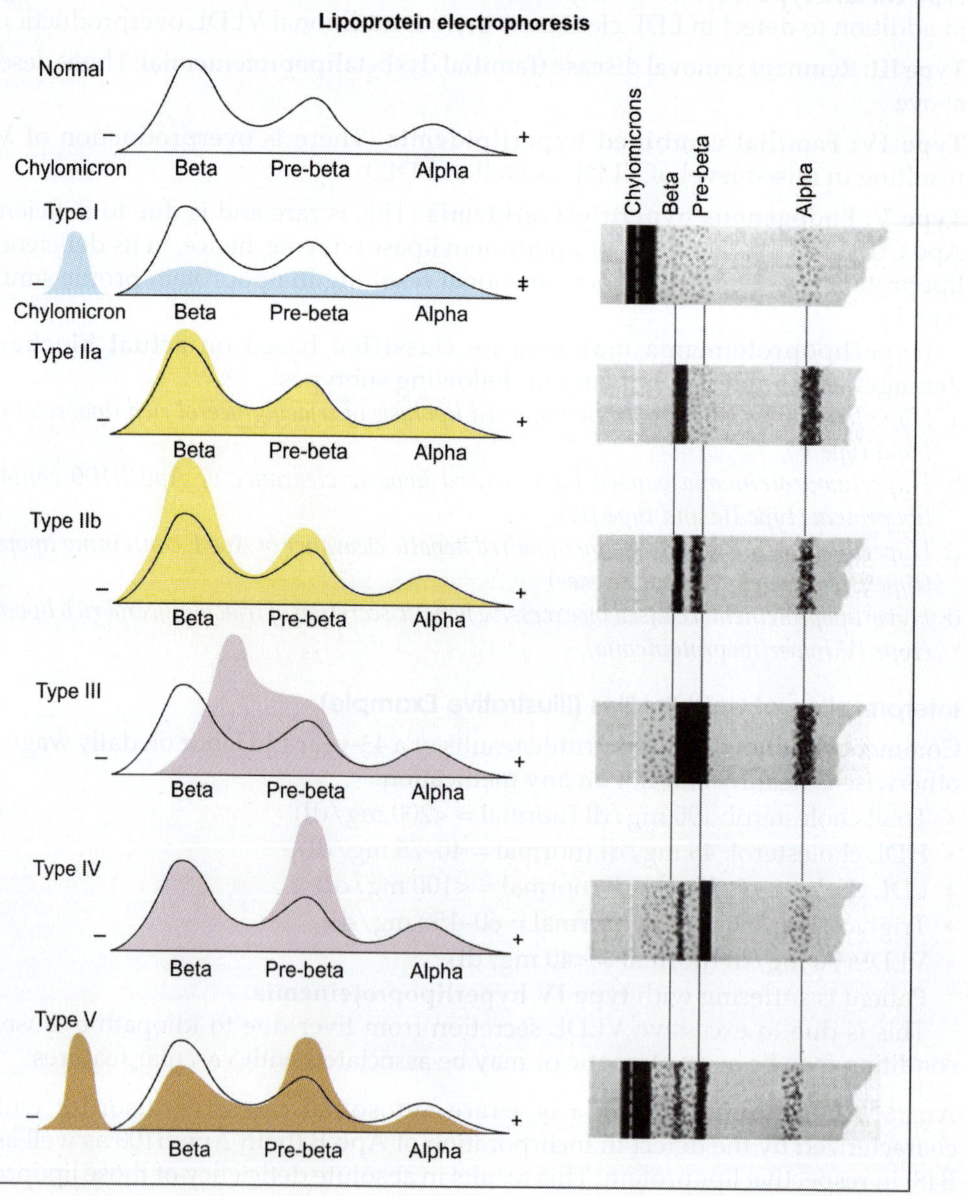

Fig. 2: Lipid profile pattern of various hyperlipidemias/ hyperlipoproteinemias

Bibliography

1. Hopkins PN, Brinton EA, Nanjee MN. Hyperlipoproteinemia Type 3: The Forgotten Phenotype. *Curr Atheroscler Rep* 2014;16:440. https://doi.org/10.1007/s11883-014-0440-2.
2. Kei A, Miltiadous G, Bairaktari E, Hadjivassiliou M, Cariolou M, Elisaf M. Dysbetalipoproteinemia: Two cases report and a diagnostic algorithm. World J Clin Cases. 2015 Apr 16;3(4):371–6. doi: 10.12998/wjcc.v3.i4.371. PMID: 25879010; PMCID: PMC4391008.

CASE

22

Xanthelasma

A 45-year-old female comes with complain of **yellow velvety soft popular lesion on inner canthus of both eyes upper eyelid** which has been present for past 2 years but has been progressively increasing in size for past 3 months. She otherwise has no other systemic complaint and is leading a normal life. She is seeking medical advice for cosmetic reasons.

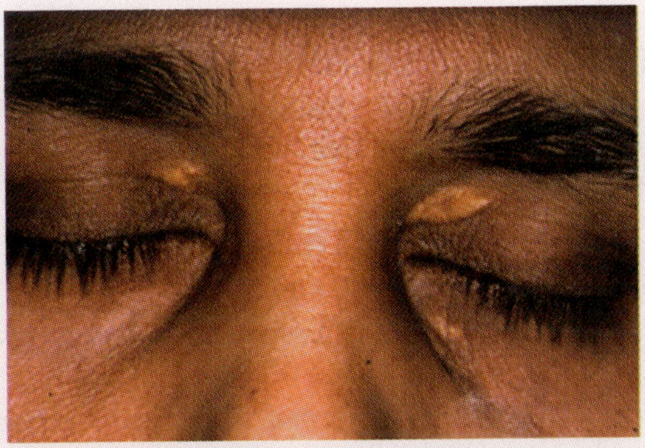

Following are the results of various laboratory investigations:

Total cholesterol: 278 mg/dl (normal = <200 mg/dl)

HDL cholesterol: 37 mg/dl (normal = 40–75 mg/dl)

LDL cholesterol (calculated): 206 mg/dl (normal = <100 mg/dl)

Triglyceride: 175 mg/dl (normal = 60–150 mg/dl)

VLDL (calculated): 35 mg/dl (normal = <40 mg/dl)

Diagnosis of xanthelasma palpebrarum is made.

QUESTIONS

Q.1. What may be reason of above presentation?

Q.2. What treatment is advised in such cases?

Q.3. What is the meaning of calculated LDL value?

Explanations

Ans.1. Xanthelasma palpebrarum is most common xanthomas. **Xanthomas** are lipid laden macrophages deposition in soft tissues. These lipids which are accumulated in macrophages are mainly cholesterol. This patient is having high cholesterol which may be reason for such deposition.

Ans.2. Since many decades, surgery has been the gold standard in the treatment of this disorder. Advent of **laser therapy (argon or carbon dioxide laser)** has revised the treatment modality in such cases and now are the preferred modality of treatment. **Cauterization** with **trichloroacetic acid** or **liquid nitrogen may be done**.

Xanthelasma is more common in female than in male. It is seen more commonly in advanced age. It is **associated with type IIa dyslipidemia** and **rarely type IIb and Type III dyslipidemia**.

Ans.3. LDL assessment may be done either by direct enzymatic method or it may be calculated using **Friedewald formula.**

Friedewald Equation

It is commonly used indirect method of LDL assessment. Total cholesterol, HDL cholesterol, triglyceride measurement are done and LDL is calculated using following formula:

LDL = (Total cholesterol) – (HDL cholesterol) – (TAG/5)

TAG/5 represents the level of VLDL cholesterol. This is because the average ratio of TAG to the cholesterol in the VLDL is five, which means TAG is almost five times to that of cholesterol in the VLDL.

Friedewald equation ***cannot*** be applied in following conditions:

1. *Samples where TAG concentration is ≥400 mg/dl*
2. *Non-fasting sample where chylomicron is also there in plasma which alters the value of TAG*
3. *Type III hyperlipoproteinemia (due to abnormal VLDL composition where TAG/5 cannot be applied)*

Bibliography

1. Dwivedi S, Aggarwal A, Singh S, Sharma V. Familial Xanthelasma with Dyslipidemia: Just Another Family Trait?. North Am J Med Sci 2012;4:238–40.
2. Manpreet Juneja, Qurratulain Chougle, Suhas Abhyankar. Rare and unusual presentation of periorbital xanthelasma. Journal of Dermatology and Dermatologic Surgery, 2016;20(2):149–51.
3. Wang P, Luo Q, Tang L. Bilateral extensive nodular xanthelasma palpebrarum: an infrequent case report. Int Ophthalmol. 2018 Apr;38(2):803–5. doi: 10.1007/s10792-017-0506-6. Epub 2017 Apr 3. PMID: 28374274.

CASE

23

Tay-Sachs Disease

An 8-month-old female child is brought to the hospital by mother with fever, vomiting and one episode of fits previous night. Mother gives the history of repeated infection to the child for which she has taken advice and treatment from a pediatrician in her colony. Baby's elder brother died at the age of 2 years due to unknown cause.

Baby is born as full term normal vaginal delivery (FTNVD) to the consanguineous parents.

On examination, baby was lethargic and had generalized hypotonia. She had bilateral nystagmus and eyes showed cherry red spot in both the eyes. Blood examination revealed decreased level of **β-hexosaminidase A enzyme**.

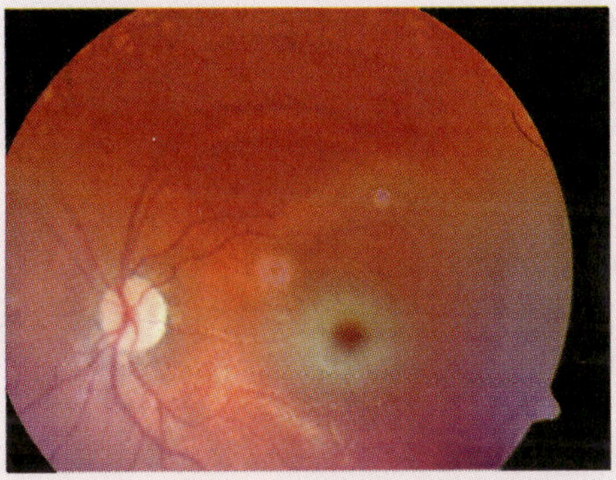

Fundus finding of cherry red spot

Based on clinical and biochemical findings, the diagnosis of Tay-Sachs disease is made.

QUESTIONS

Q.1. What is the etiopathogenesis of Tay-Sachs disease?

Q.2. How many subtypes of Tay-Sachs disease are described?

Q.3. Why there is cherry red spot in retina? What are other conditions in which baby has cherry red spot in the retina?

Q.4. What is the current modality of treatment of this disease?

Explanations

Ans.1. **Tay-Sachs disease (GM2 gangliosidosis type I)** an **autosomal recessive lysosomal storage disorder** which involves central nervous system. It is due to mutation of **gene of β-hexosaminidase A enzyme.** Gene for this enzyme is located on chromosome 15q23. Deficiency of this enzyme results in defect of degradation of ganglioside, it results in accumulation of **GM2 trihexosylceramide.** Predominant cells where such accumulation take place are neurocytes and retinal ganglions.Tay-Sachs disease was described by **Tay in 1881 and by Sachs in 1887.** It is one of most widely studied LSDs.

Prevalence is 1/200000 in general population with very high prevalence in Ashkenazi Jews at the rate of 1/4000.

Ans.2. **Subtypes of Tay-Sachs disease** are:
- Infantile
- Juvenile
- Late onset adult type

Ans.3. **Cherry red spot** was first described by Warren Tay in 1881. Cherry red spot is due to accumulation of **sphingolipids in retinal ganglion cell giving the fundus the chalky white appearance.**

Many other conditions have this characteristic ophthalomoscopic finding of cherry red spot. They are:
- Sandhoff disease (GM2 type II)
- Gangliosidosis GM2 type III and GM1 type I
- Niemann-Pick disease
- Sialidosis types I and II
- Farber disease
- Mucolipidosis III
- Metachromatic leukodystrophy
- Multiple sulfatase deficiency
- Toxicity (dapsone)
- Wolman disease
- Central retinal artery occlusion.

Ans.4. **Enzyme replacement therapy (ERT)** and gene therapy are being tried in treatment of such patients. Neurological involvement is irreversible and only symptomatic management is possible for such involvements.

Bibliography

1. Aragão Ricardo Evangelista Marrocos de, Ramos Régia Maria Gondim, Pereira Felipe Bezerra Alves, Bezerra Andreya Ferreira Rodrigues, Fernandes Daniel Nogueira. "Cherry red spot" in a patient with Tay-Sachs disease: case report. Arq. Bras. Oftalmol. [Internet]. 2009 Aug [cited 2020 Nov 25]; 72(4): 537–539.
2. Arisoy AE, Ozden S, Ciliv G, Ozalp I. Tay-Sachs disease: a case report. Turk J Pediatr. 1995 Jan-Mar;37(1):51–6. PMID: 7732608.

CASE

24

Niemann-Pick Disease

A 5-year-old male boy born to consanguineous parents is brought to pediatric OPD by mother with complain of distension of abdomen and failure to grow and history of delayed milestone. He has difficulty in walking with repeated episode of fall and an episode of generalized tonic clonic convulsion 2 months back.

Examination revealed hepatosplenomegaly.

Bone marrow biopsy revealed lipid laden foam histiocyte known as Niemann-Pick cell.

Enzyme assay revealed deficiency of enzyme sphingomyelinase in WBC.

Diagnosis of Niemann-Pick disease is made.

QUESTIONS

Q.1. What is the cause of this disease?

Q.2. How many subtypes of Niemann-Pick disease are described?

Q.3. How to make diagnosis in such cases?

Q.4. What is the treatment advised?

Explanations

Ans.1. **Niemann-Pick disease** is an **autosomal recessive** condition which is due to **deficiency of acid sphingomyelinase** in the lysosome resulting in accumulation of sphingomyelin and cholesterol in lysosome of reticuloendothelial cell of liver, spleen, bone marrow, brain and lung.

In Fig. 1, the plan of sphingolipid breakdown is described. Table 1 summarises the respective defect and corresponding material accumulated in various diseases.

Table 1: Various sphingolipidosis

Disease	*Enzyme deficiency*	*Material accumulated*
Tay-Sachs disease	Hexosaminidase A	GM2 ganglioside
Niemann-Pick disease	Sphingomyelinase	Sphingomyelin
Gaucher's disease	Glucocerebrosidase	Glucocerebroside

Ans.2. Subtypes of Niemann-Pick disease are **A, B, C and D**. In types A and B, there is deficiency of acid sphingomyelinase which helps to breakdown sphingomyelins,

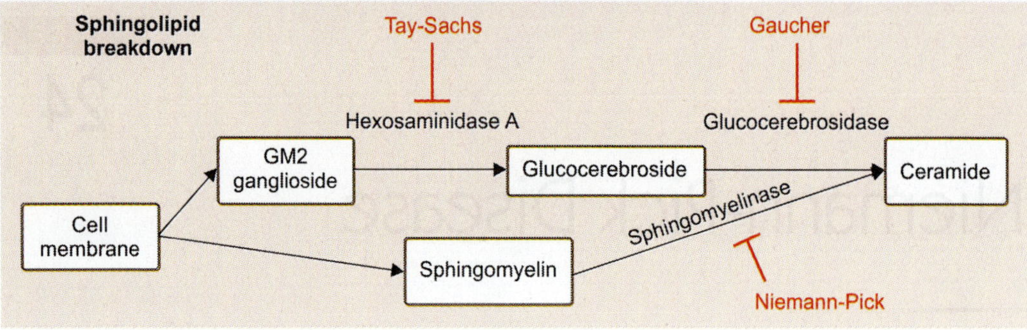

Fig. 1: Plan of sphingolipid degradation and defect in various sphingolipidoses

whereas types C and D are caused by a defect in the transport of intracellular cholesterol.

Types A and B are included in type I and **types C and D are included in type II**. Type I is more severe form than type II. Epidemiological studies have led to the definition of early infantile (at <2 years of age), late infantile (at age 2 to <6 years), juvenile (at age 6 to <15 years) and adolescent/adult-onset (≥15 years) forms of the disease.

Ans.3. Niemann-Pick disease is characterized by **failure to thrive, hepatosplenomegaly** and **neurodegenerative** changes. Organomegaly is an important finding in NP disease. As mentioned in previous case with discussion of cherry red spot, NP disease also **cherry red spot** may be found. Low enzyme activity (<10% of normal) may be detected in nucleated cell (WBC or cultured fibroblast).

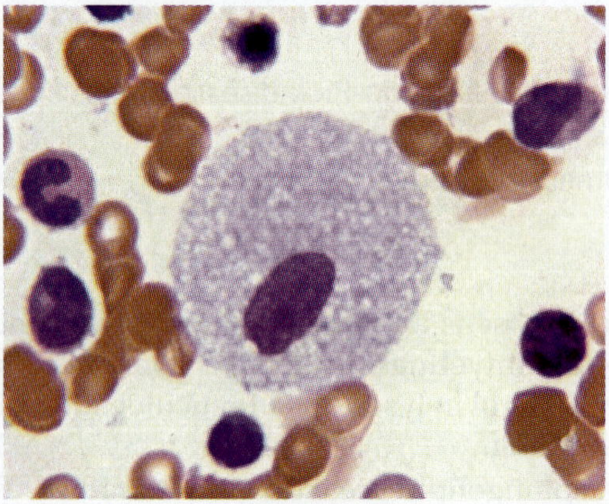

Fig. 2: Bone marrow involvement in NP disease

Bone marrow biopsy may be done which reveals Niemann-Pick cell.

Ans.4. No definitive treatment available. Enzyme replacement therapy (ERT) is in phase 3 trial. Genetic counseling to be done.

Bibliography

1. Abe K, Sakai N (2017). Patient with Niemann-Pick disease type C: over 20 years' follow-up. *BMJ case reports, 2017*, bcr2017220134. https://doi.org/10.1136/bcr-2017-220134.

CASE

25

Jamaican Vomiting Sickness

A 57-year-old native of West Africa comes to emergency unit with history of severe vomiting, epigastric discomfort and confusion for past 8 hours. He is a known diabetic for past 2 years and has been taking some local herbs prevalent at his native for lowering the blood sugar for past 6 months. He has visited India to meet his relative. He is not sure if he has consumed excess of that local herbal medicine mistakenly.

On examination, he was comatose and dehydrated.

Heart rate: 110 /min (tachycardia)

Respiratory rate: 34/min (tachypnea)

Blood investigation revealed

Blood glucose: 43 mg/dl (normal random = 80–140 mg/dl)

Serum electrolytes:

- Na^+ = 124 mEq/L (normal = 135–145 mEq/L)
- K^+ = 5.1 mEq/L (normal = 3.5–5.0 mEq/L)
- Cl^- = 90 mEq/L (normal = 90–110 mEq/L)

Urine showed excessive amount of dicarboxylic acid.

Patient was immediately infused with 10% dextrose and electrolyte imbalance was treated.

QUESTIONS

Q.1. What diagnosis is suspected in above case? Explain the toxin involved.

Q.2. What is the reason of low blood glucose?

Q.3. Comment on outbreak of *Litchi* fruit poisoning in Bihar, India in 2017.

Explanations

Ans.1. This patient seems to have **"Jamaican vomiting sickness"**, a disease caused by consumption of **unripe fruit of akee tree.** Unripe fruit of Akee tree **(*Blighia sapida*)** contains a toxin known as **hypoglycin** which is metabolized in the liver by deamination and oxidative decarboxylation to produce **methylenecyclopropyl acetyl-CoA (MCPA).**

MCPA inhibits medium chain acyl-CoA dehydrogenase thus inhibiting fatty acid beta oxidation of long chain fatty acid at butyryl-CoA dehydrogenase step.

The ripe fruit flesh of akee contains only low quantities of hypoglycins, but concentrations in unripe fruits are 10–100 times greater amount and hence very toxic.

Akee (*Blighia sapida*) is a commonly eaten fruit that is indigenous to **West Africa and Jamaica**. It is used as medication to treat diabetes because of its tendency to lower the blood glucose.

Ans.2. **Reason for hypoglycemia in akee fruit poisoning.**

Lack of beta oxidation causes excess utilization of blood glucose resulting in low blood glucose. This hypoglycemia is not compensated by gluconeogenesis as gluconeogenesis process require NADH from uninterrupted fatty acid beta oxidation. Hence, in toxicity by akee fruit, patient develops severe hypoglycemia and end up in coma and death if medical attention is not given timely.

Ans.3. In year **2017,** hundreds of children died in city **Muzaffarpur, Bihar, India,** after consuming excess amount of **Litchi fruit,** which brought world attention to toxicity of excess consumption of this fruit. Children were all from **poor farmer's family** who could not afford evening meal, forcing them to consume excess amount of **Litchi** which was cultivated in their farm.

Fig. 1: Litchi fruits

This illness was characterized by acute seizure and changed mental status starting in early morning and associated with high mortality. **Hypoglycemia** was the characteristic finding in all these children.

Clinical presentation was similar to Jamaican vomiting sickness reported in various parts of the world.

Later the toxin **methylenecyclopropyl glycine (MCPG)** and **hypoglycin A** was isolated from **fruit of Litchi** which were responsible for hypoglycemic attack (specially the raw fruit).

Bibliography

1. Shrivastava A, et al. Association of acute toxic encephalopathy with Litchi consumption in an outbreak in Muzaffarpur, India, 2014: A case-control study. Lancet Glob Health. 2017 Apr;5(4): e458-e466. doi: 10.1016/S2214-109X(17)30035-9. Epub 2017 Jan 31. PMID: 28153514.
2. Tanaka K, Kean EA, Johnson B. Jamaican vomiting sickness. Biochemical investigation of two cases. N Engl J Med. 1976 Aug 26;295(9):461-7. doi: 10.1056/NEJM197608262950901. PMID: 940578.

CASE 27

Urea Cycle Disorder

A 2-year-old male baby is presented with history of frequent episode of vomiting, lethargy*, restlessness and aversion to protein containing food. On examination, baby was lethargic and mentally retarded.

Following are the results of various laboratory investigations:

Ammonia level in the blood 567 µmol/L (normal is 25 to 100 µmol/L)

Citrulline level in blood was low and ornithine was high. Glutamine was found in excess in blood.

Urine shows excretion of orotic acid. Orotic acid and creatinine ratio was 5.7 mmol/mol (normal is 1.0 to 3.5 mmol/mol)

Diagnosis of urea cycle disorder is made.

QUESTIONS

Q.1. What may be the most probable enzyme deficiency in this baby? How many types of urea cycle disorders are known?

Q.2. What is orotic acid and what may be the cause of orotic aciduria in this baby?

Q.3. Explain the biochemical reason for clinical signs and symptoms in this baby?

Q.4. What is the line of management for urea cycle disorder?

Explanations

Ans.1. Low level of citrulline and high level of **ornithine** suggest deficiency of enzyme **ornithine transcarbamylase (OTC).**

OTC is the enzyme which catalyzes second step of urea cycle converting ornithine and **carbamoyl phosphate** to **citrulline** (Fig. 1), hence, in deficiency of this enzyme, level of citrulline will decrease with simultaneous increase of ornithine.

OTC gene is located at **Xp21.1**. OTC deficiency is an **X-linked recessive disorder** and males are more seriously affected than heterozygous female. Rarely this mutation is sporadic and may appear later in life when the presentation is late.

OTC enzyme deficiency is the most common cause of urea cycle disorder responsible for almost 50% cases of urea cycle disorder.

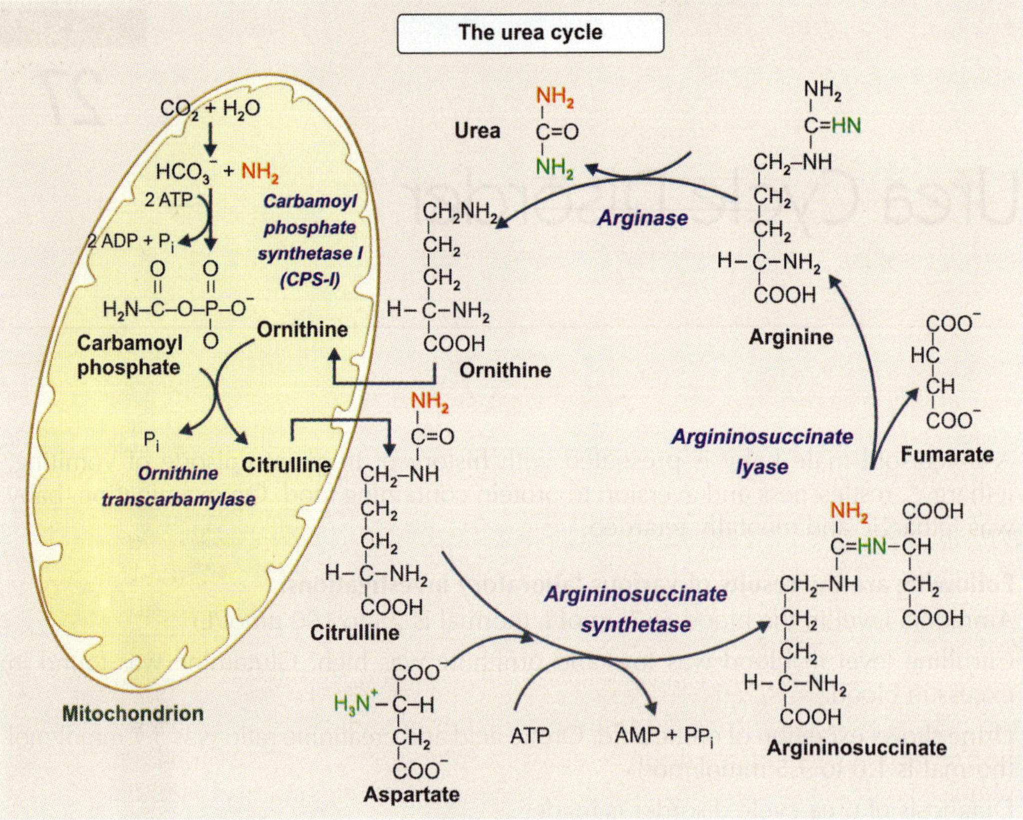

Fig. 1: Steps involved in urea cycle

Various types of urea cycle disorder are summarized in given table below.

Enzyme deficiency	*Disorder*	*Inheritance*	*Clinical presentation*
CPS I deficiency	**Hyperammonemia**	AR	Raised ammonia in blood Lethargy, coma, aversion to protein containing food.
OTC deficiency	**Hyperammonemia type II**	**XR****	Raised ammonia in blood Lethargy, coma, aversion to protein containing food.
Ornithine carrier (ORNT1) defect	Hyperammonemia **Hyperornithinemia** **Homocitrullinemia**	AR	Raised ammonia in blood Lethargy, coma, aversion to protein containing food.
Argininosuccinate synthetase deficiency	**Citrullinemia type I**	AR	Raised ammonia in blood Lethargy, coma, aversion to protein containing food, liver failure
Citrin* deficiency	**Citrullinemia type II**	AR	Jaundice in neonate, stupor, coma in adults

(Contd.)

Enzyme deficiency	*Disorder*	*Inheritance*	*Clinical presentation*
Argininosuccinate lyase deficiency	**Argininosuccinic aciduria/ argininosuccinic urea**	AR	Raised ammonia in blood Lethargy, coma, aversion to protein containing food. Trichorrhexis nodosa
Arginase	Hyperargininemia	AR	Microcephaly, low IQ, spastic tetraparesis
NAG synthetase deficiency	Very rare	AR	Raised ammonia in blood Lethargy, coma, aversion to protein containing food.

*__Citrin__, a mitochondrial carrier protein of glutamate and aspartate **(CTLN2)**

**All urea cycle disorders are AR except OTC deficiency which is XR.

Ans.2. **Orotic acid is the intermediate of pyrimidine nucleotide biosynthesis.**

Increased level of orotic acid in urine suggests the deficiency of OTC. Deficiency of OTC causes accumulation of carbamoyl phosphate in the mitochondria. Excess level of carbamoyl phosphate shunts from mitochondria to cytosol and becomes the intermediate of pyrimidine nucleotide biosynthesis. This results in production of excessive orotic acid which is excreted in the urine.

Ans.3. On consumption of protein-rich food, symptoms of ammonia toxicity precipitate and baby slowly learn to avoid protein containing food. Excess ammonia is toxic for the brain and may even results in cerebral atrophy, edema, seizures and death. Mental retardation is due to excess ammonia and delay in seeking medical attention timely.

Ans.4. **Sodium benzoate, sodium phenylacetate** is given for therapy to get rid of source of ammonia. **Sodium benzoate combines with glycine** and is excreted as **hippuric acid** and **phenylacetate combines with glutamine** and is excreted as **phenylacetylglutamine** which is also excreted in the urine.

This along with restriction of protein and supplementation of citrulline and arginine (except in case of hyperargininemia) is shown to have promising result and improves the general condition of the patient.

Liver transplantation in extreme cases may be planned.

Bibliography

1. Foschi FG, Morelli MC, Savini S, Dall'Aglio AC, Lanzi A, Cescon M, Ercolani G, Cucchetti A, Pinna AD, Stefanini GF. Urea cycle disorders: a case report of a successful treatment with liver transplant and a literature review. World J Gastroenterol. 2015 Apr 7;21(13):4063–8. doi: 10.3748/wjg.v21.i13.4063. PMID: 25852294; PMCID: PMC4385556.
2. Gallagher MM, Clancy GT. A case study: urea cycle disorder. Neonatal Netw. 1991 Sep;10(2):35–44. PMID: 1886557.
3. Wang B, Jha P. A Case of Atypical Adult Presentation of Urea Cycle Disorder. WMJ. 2019 Jul;118(2):98–100. PMID: 31532938.

Word Meaning

Lethargy: Lack of energy and enthusiasm.

CASE

28

Phenylketonuria (PKU)

A 2-year-old child presents to hospital with history of generalized fits. Baby was born full term by normal vaginal delivery (NVD) without forceps application. On examination, baby had delayed milestone and was lethargic, **hair was grey** colored and **skin was light colored**. Mother gave the history of **mousy odor in the urine.**

QUESTIONS

Q.1. What is the most probable diagnosis in this case? What is the enzyme deficiency?

Q.2. Why the baby's hair is grey and skin is light colored?

Q.3. Why the odor of urine is **mousy odor**? What is the other name given to the odor?

Q.4. What is **malignant PKU**?

Q.5. What test is done to confirm the diagnosis?

Q.6. What treatment is advised for this baby?

Q.7. What is the challenges in management of PKU patients in Indian scenario?

Explanations

Ans.1. This baby is suffering with **phenylketonuria (PKU)** which is due to defect in enzyme/s responsible for conversion of phenylalanine to tyrosine. In a normal metabolic pathway, phenylalanine is converted to tyrosine by hydroxylation of benzene ring in the side chain of phenylalanine with the help of enzyme **phenylalanine hydroxylase** (PAH).

Defect in conversion of phenylalanine to tyrosine results in alternate pathway for utilization of phenylalanine. In alternate pathway, phenylalanine is converted to **phenylketones** which cause impaired neurological development and mental retardation.

Gene of PAH is found on chromosome 12 at 12q23-2 location. This gene has 13 exons and more than 500 mutations are known in it.

Incidence of PKU is highest in Caucasians with 1 in 10,000 live birth. In many countries, there is provision of newborn screening programme. PKU is one of the diseases being screened in NSP.

CASE

29

Albinism

An 18-year-old male presents to ophthalmology OPD with complain of difficulty in visualising far objects and involuntary movements in both the eyes since long. He also complains of skin irritation and redness on exposed skin when he go out in sun. His skin is white and hair is light brown. Both the eyes show involuntary movements **(nystagmus)**.

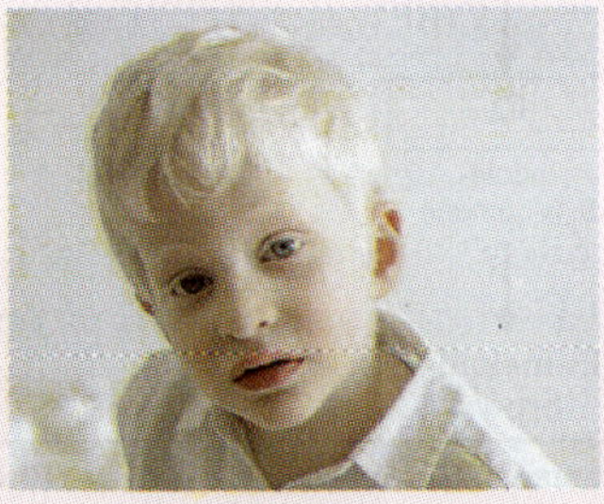

Albinism

QUESTIONS

Q.1. What is the condition this male is suffering from? What is the biochemical pathway involved in this disease?

Q.2. What may be the complication of this disease?

Q.3. What is the line of management in this case?

Explanations

Ans.1. This male is suffering with **oculocutaneous albinism (OCA)**. It is an **autosomal recessive** disorder characterized by **mutation of tyrosinase gene.** Gene for tyrosinase enzyme is located on **chromosome 11q14.3.** Gene consists of 5′ exon and code for tyrosinase enzyme which is 529 amino acid long protein. This enzyme catalyzes initial two steps of melanin synthesis. **Conversion of tyrosine to L-DOPA and then L-DOPA to DOPAquinone.**

Lack of tyrosinase enzyme results in deficiency or absence of pigment melanin from skin, iris, and hair.

1 in 20,000 individuals worldwide are affected with albinism with racial difference in prevalence of this disease.

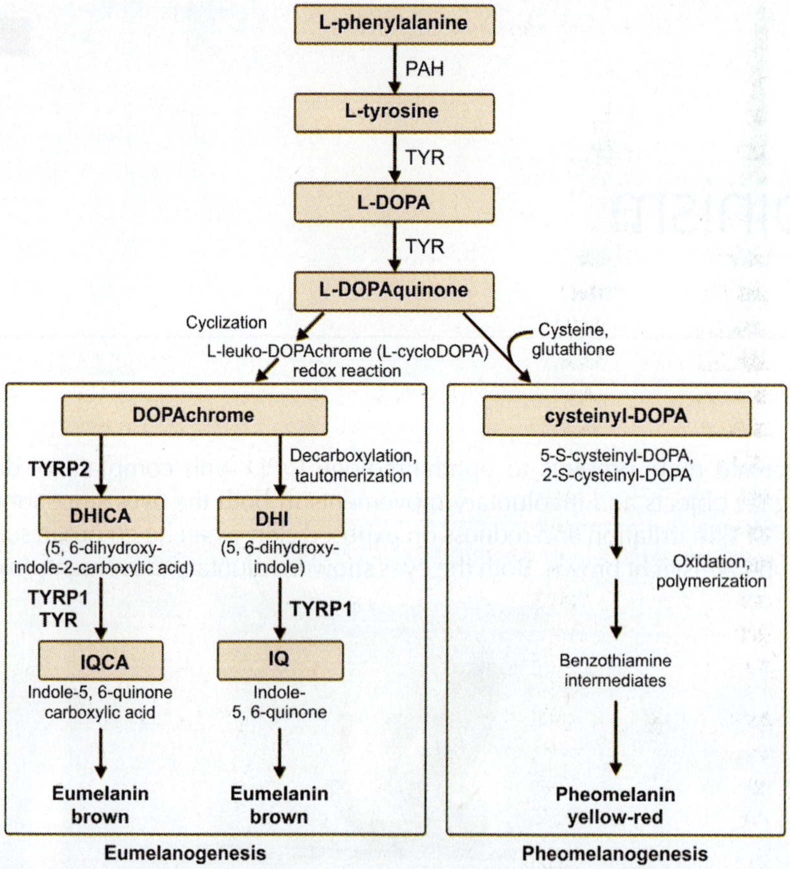

Fig. 1: Synthesis of eumelanin and pheomelanin from tyrosine

Based on clinical spectrum, oculocutaneous albinism **(OCA) is of four types, OCA1, OCA2, OCA3, OCA4.** OCA1 has further two subtypes, OCA1A and OCA1B. OCA1 A+B is most common and most severe type characterized by complete absence of melanin throughout life.

Ans.2. Patient wih albinism may have **photophobia, strabismus, nystagmus, decreased visual acuity.** Oculocutaneous albinism (OCA) has photosensitivity and patient tends to develop erythema and blistering on exposure to sunlight.

Albinos are more prone for nonmelanomatous skin cancer-like squamous cell and basal cell cancer.

Ans.3. Care for the skin by applying sunscreen before sunexposure and care for the eye is advised.

Bibliography

1. de Gennaro G, Vitti P, Marinò M. A Case of Oculocutaneous Albinism in a Patient with Hashimoto's Thyroiditis. Eur Thyroid J. 2019 Jan;8(1):56–58. doi: 10.1159/000493730. Epub 2018 Oct 16. PMID: 30800642; PMCID: PMC6381916.
2. Karaman A. Oculocutaneous albinism type 1A: a case report. Dermatol Online J. 2008 Nov 15;14(11):13. PMID: 19094851.

CASE 31

Pheochromocytoma

A 34-year-old female reports to emergency with history of severe frontal headache and blurring of vision, palpitation* and excessive sweating on repeated occasions for past 6 months.

On examination, her BP was recorded as 168/110 mm Hg.

Serum showed high catecholamine and urine showed high level of metanephrine and normetanephrine.

Serum catecholamines (ng/L)	Value	Normal level
Total	22326	<598
Norepinephrine	20591	<420
Epinephrine	1318	<84
Urinary metanephrines (μg/24-hr)		
Metanephrines	2570	<90
Normetanephrines	**>4800**	**<180**

She was suspected of pheochromocytoma and CT abdomen was advised.

CT revealed a 4 cm × 3 cm × 4.7 cm nodular mass on right side adrenal gland.

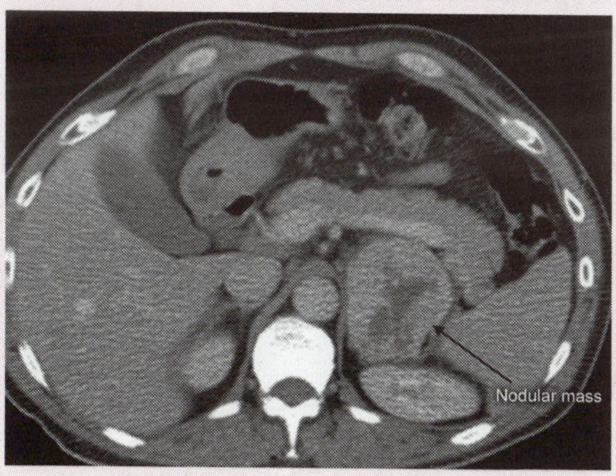

Nodular mass lesion on right adrenal gland suggestive of pheochromocytoma

QUESTIONS

Q.1. What may be the possible diagnosis?

Q.2. How the diagnosis is confirmed?

Q.3. What is the classic "rule of ten" for pheochromocytomas?

Q.4. What is the treatment in this disorder?

Explanations

Ans.1. This patient is having a rare tumor **pheochromocytoma** which produces **catecholamines.** It **originates in chromaffin cell** and is mostly located at adrenal gland medulla.

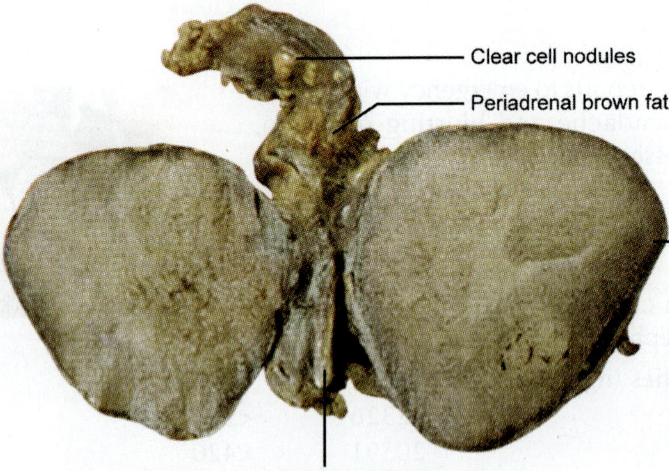

Fig. 1: Tumor of adrenal gland

Classical triad of symptoms of pheochromocytoma is palpitation, episodic headache and excess sweating (sudoresis). The triad of symptoms along with hypertensions makes the diagnosis of pheochromocytoma most likely.

Occurrence of pheochromocytoma may be sporadic or it may be associated with **multiple endocrine neoplasia 2 (MEN2), VHL (von Hippel-Lindau) disease.** It occurs with equal frequency in male and female.

Ans.2. Blood and urine investigations for estimation of **catecholamine** and their methylated metabolite **metanephrine** is to be done along with abdominal CT scan. Most pheochromocytomas secrete norepinephrine predominantly. Normal adrenal medulla secretes epinephrine predominantly. Catecholamines (norepinephrine, epinephrine and dopamine) can be measured by HPLC, ELISA, LC/MS.

Ans.3. *Classic rule of ten is:*

- 10% is bilateral
- 10% is extra adrenal
- 10% is malignant

Ans.4. Definitive treatment is surgical where complete tumor removal is the target by partial or total **adrenalectomy.**

Bibliography

1. Alface MM, Moniz P, Jesus S, Fonseca C. Pheochromocytoma: clinical review based on a rare case in adolescence. BMJ Case Rep. 2015 Aug 4;2015:bcr2015211184. doi: 10.1136/bcr-2015-211184. PMID: 26243750; PMCID: PMC4533612.
2. Molaei A, Abarzadeh-Bairami V and Sadat-Ebrahimi SR. A case of pheochromocytoma presenting with cardiac manifestation: case report. BMC Pediatr 2020;20:299.

Word Meaning

Palpitation: Rapid, irregular and strong heartbeat which is noticeable by the patient.

CASE

32

Parkinson's Disease

A 72-year-old male comes to OPD with complain of intermittent involuntary tremors in right hand for past 6 months which has been progressively getting worsened over past 1 month. He finds difficulty in feeding himself and often spills food on dining table. He finds difficulty in walking and coordination of gait. On examination, his cognitive function was normal and he had slow and shuffling gait. His hands showed resting tremors.

Diagnosis of Parkinson disease is made and he is prescribed levodopa.

QUESTIONS

Q.1. Comment on the diagnosis and explain its biochemical basis.

Q.2. What is the rationale behind giving levodopa? What other treatment strategies are adopted for treating this disorder?

Explanations

Ans.1. **Parkinson disease** is a disorder of old age which manifests in old age beyond 60 years, though early onset variant is also known where disease onset occurs around 4th decade of life. Males are predominantly affected than females.

This disease is characterized by **resting tremor, rigidity, bradykinesia*** and **postural instability.** Patient often encounters difficulty in walking with frequent falls which begin gradually and get worse over time. Other features include mental and behavioral changes, sleep problems, depression, memory difficulties, and fatigue (Fig. 1).

This disease is due to loss of neurons in the brain which are responsible for production of dopamine.

Dopamine is a neurotransmitter which is synthesized from tyrosine in following pathway (Fig. 2).

Reason for lack of dopamine in Parkinson's disease is neuronal degeneration.

Ans.2. Levodopa enters the brain and acts as precursor of dopamine and relieves the symptom of disease. **Dopamine agonists dihydroergocryptine, pergolide, pramipexole** may also be advised.

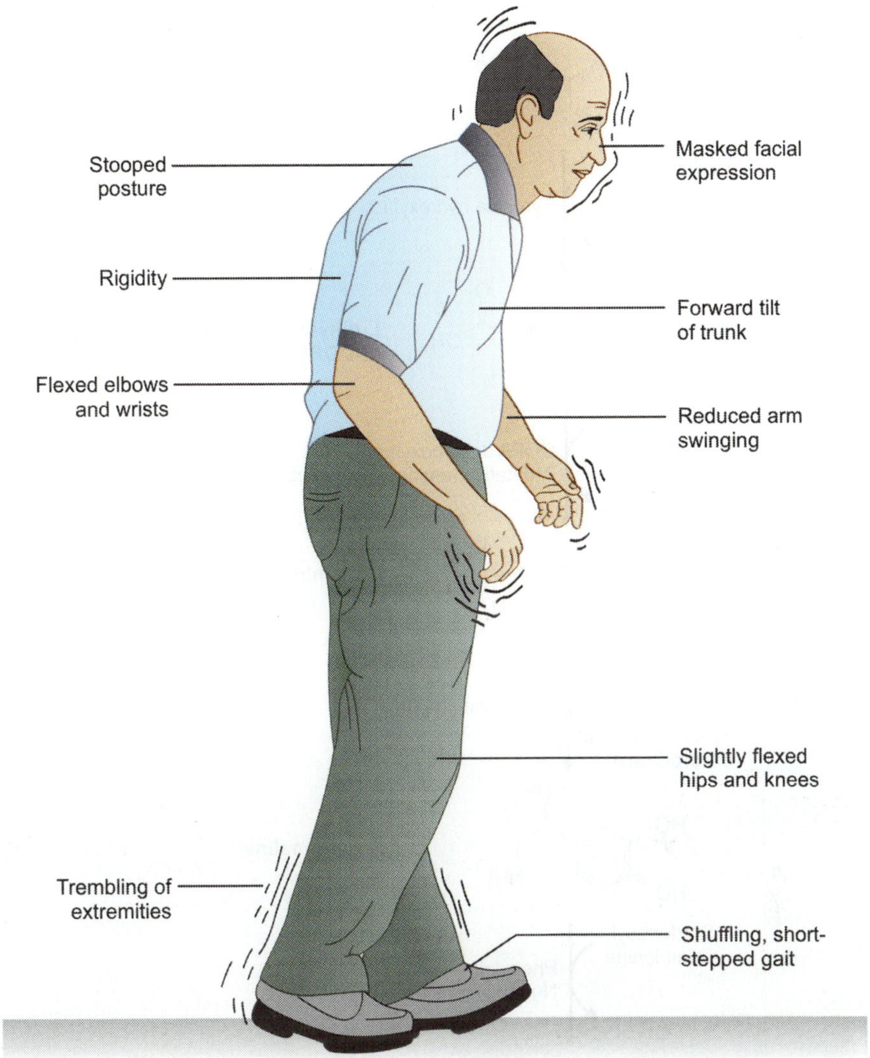

Fig. 1: Characteristic features of Parkinson disease

Treatment objectives include the management of motor complications, the prevention of falls, the treatment of psychosis and dementia.

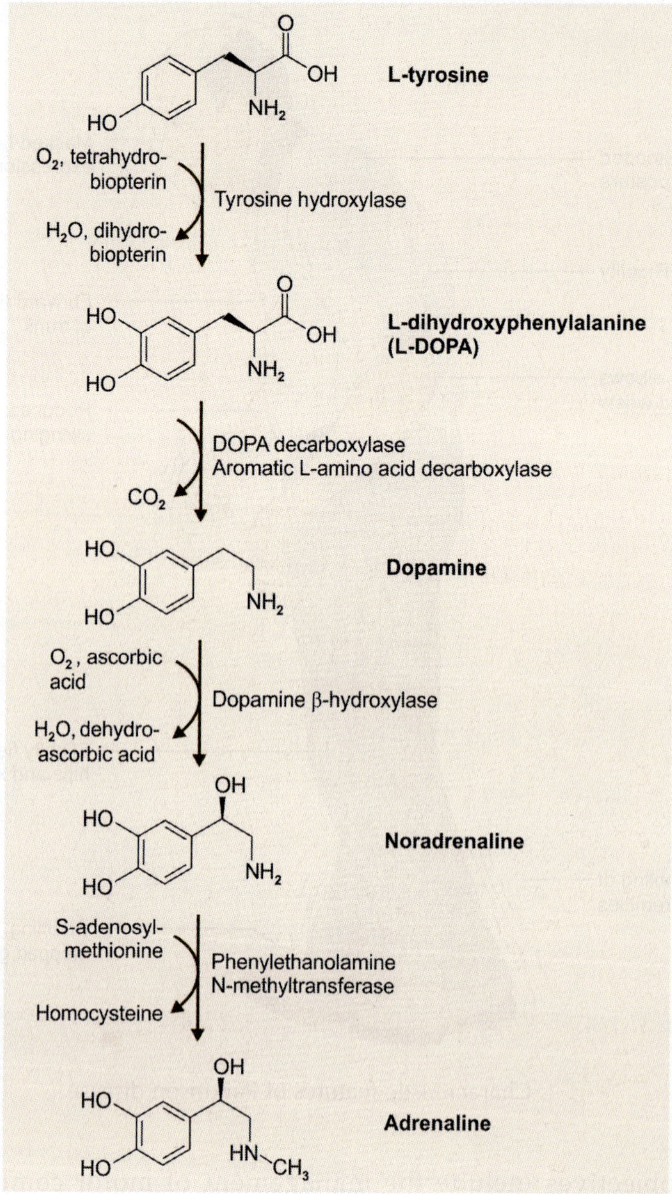

Fig. 2: Pathway of dopamine synthesis from tyrosine

Bibliography

1. Sekeff-SallemFlávio Augusto, Barbosa Egberto Reis. Diagnostic pitfalls in Parkinson's disease: case report. Arq. Neuro-Psiquiatr. [Internet]. 2007 June [cited 2021 Jan 17; 65(2A):348–51.
2. Zou K, Guo W, Tang G, Zheng B, Zheng Z. A Case of early onset Parkinson's disease after major stress. Neuropsychiatr Dis Treat. 2013;9:1067–9. doi: 10.2147/NDT.S48455. Epub 2013 Aug 2. PMID: 23950648; PMCID: PMC3742351.

Word Meaning

Bradykinesia: Slow/sluggish movement.

CASE

33

Tyrosinemia Type I (Tyrosinosis)

A 2-month-old female is presented with complain of failure to gain weight, frequent episode of diarrhea and vomiting since birth. Her skin is pale and yellow discoloration of eye was found by attending pediatrician. Hepatomegaly was found and renal parameters were deranged.

Urine had **cabbage odor/rancid butter odor** and it showed the presence of **succinylacetone.** Defect of amino acid metabolism was suspected and patient was prescribed **nitisinone.**

QUESTIONS

Q.1. What may be the diagnosis in this child?

Q.2. What biochemical pathway is involved in this disorder? Mark the step involved in this disease.

Q.3. Enumerate other metabolic errors encountered in this pathway along with respective enzyme deficient.

or

Name the other disorders associated with the concerned pathway involved in the given disorder in this case study.

Explanations

Ans.1. **Cabbage odor/rancid butter odor** and presence of **succinylacetone** in the urine is the characteristic finding seen in **tyrosinemia type I**, a disorder of tyrosine catabolic pathway. This is also known as **hepatorenal tyrosinemia** as it involved liver and kidney predominantly as seen in above illustrative case as well. Presentation of this disorder in early (first three months of life), and baby commonly presents with **failure to thrive** and hepatorenal involvement.

Ans.2. The pathway involved is **tyrosine catabolism** and the enzyme deficient is **fumarylacetoacetate hydrolase (FAH).** This enzyme is required at last step of tyrosine catabolism to finally degrade the **fumarylacetoacetate to fumaric acid and acetoacetic acid.** In deficiency of this enzyme, maleylacetoacetate and fumarylacetoacetate are converted to **succinylacetoacetate** which then is converted to **succinylacetone** and gets excreted in urine.

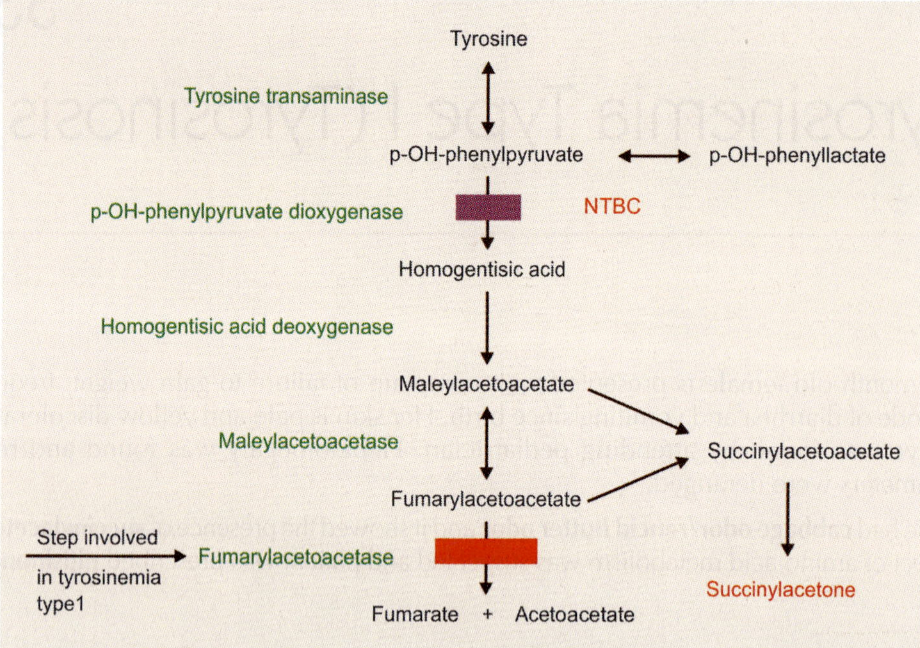

Fig. 1: Pathway of tyrosine degradation and step involved in tyrosinemia type I. Also shown is the step of **nitisinone (NTBC)** in blocking the earlier step of the pathway

Decreased activity of **FAH enzyme** is assessed in fibroblast.

Nitisinone is prescribed for treatment which blocks the pathway at early step and prevents the accumulation of toxic intermediates. In addition, low tyrosine diet is advised.

Ans.3. Other metabolic errors of tyrosine catabolism with respective enzyme deficiency are enumerated in Table 1.

Table 1: Disorders involved in tyrosine degradation

Disorder	*Enzyme deficiency*
Tyrosinemia type II (oculocutaneous tyrosinemia, also known as Richner-Hanhart syndrome)	**Tyrosine transaminase**
Tyrosinemia type III (neonatal tyrosinemia)	**Parahydroxyphenyl pyruvic acid dioxygenase**
Alkaptonuria	**Homogentisic acid oxidase**

Bibliography

1. Rashad MM, Nassar C. Tyrosinemia TypeI: A case report. Sudan J Paediatr. 2011;11(1):64–7. PMID: 27493308, PMCID: PMC4949785.
2. Uçar HK, Tümgör G, Kör D, Karda? F, Mungan NO. A Case Report of a Very Rare Association of Tyrosinemia type I and Pancreatitis Mimicking Neurologic Crisis of Tyrosinemia Type I. Balkan Med J. 2016 May;33(3):370–2. doi: 10.5152/balkanmedj.2016.141074. Epub 2016 May 1. PMID: 27308087; PMCID: PMC4899002.

CASE 34

Tyrosinemia Type II (Richner-Hanhart Syndrome/ Oculocutaneous Involvement)

A 2-year-old male child presents with ulceration in eye and thick rough patchy skin on palm and foot for past 6 months.

On examination, eye revealed dendritic corneal ulceration and skin of palm and plantar aspects of foot showed keratosis.

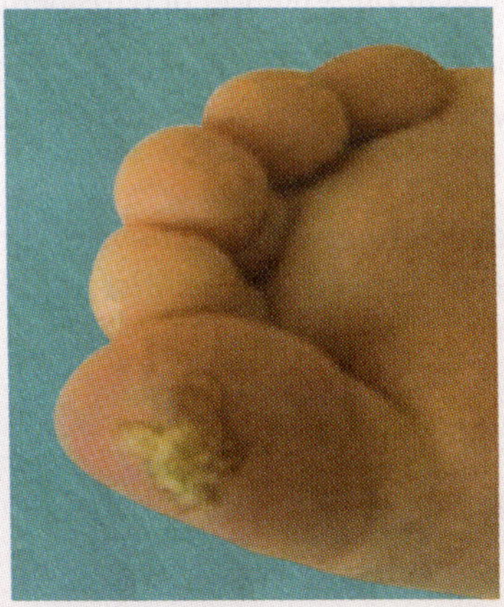

Hyperkeratosis of plantar aspect of great toe

Blood and urine examination showed raised level of tyrosine amino acid.

Corneal ulcerations were not responding to antiviral treatment.

Amino acid metabolic error was suspected based on above biochemical and clinical findings.

QUESTIONS

Q.1. What may be the diagnosis in above case?

Q.2. What is the step involved in this disorder?

Explanations

Ans.1. Raised tyrosine level in biological fluid along with characteristic dendritic ulceration of cornea and keratosis of palmoplantar skin suggests the diagnosis of oculocutaneous tyrosinemia known as ***tyrosinemia type II or Richner-Hanhart syndrome.***

Characteristic clinical features of this disorder are depicted in Fig. 1.

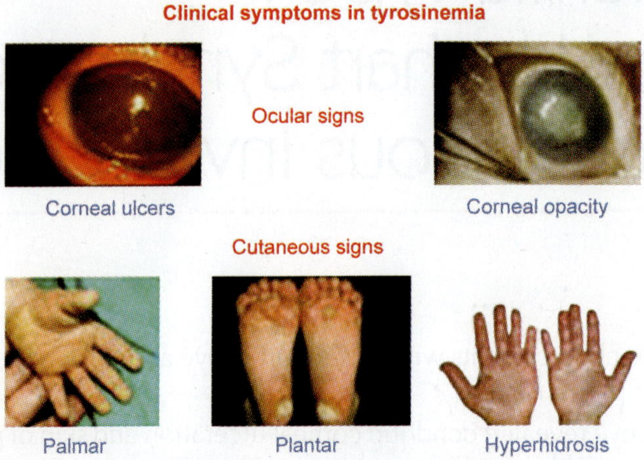

Fig. 1: Clinical presentation of tyrosinemia type II

This disease must be suspected in all cases of dendritic keratitis not reacting on the antiviral treatment, specially if associated with cutaneous lesions as described in this case.

Ans.2. Tyrosinemia type II or Richner-Hanhart syndrome is an autosomal recessive disorder which is due to tyrosine transaminase deficiency. The steps involved in various types of tyrosinemia are depicted in Fig. 2.

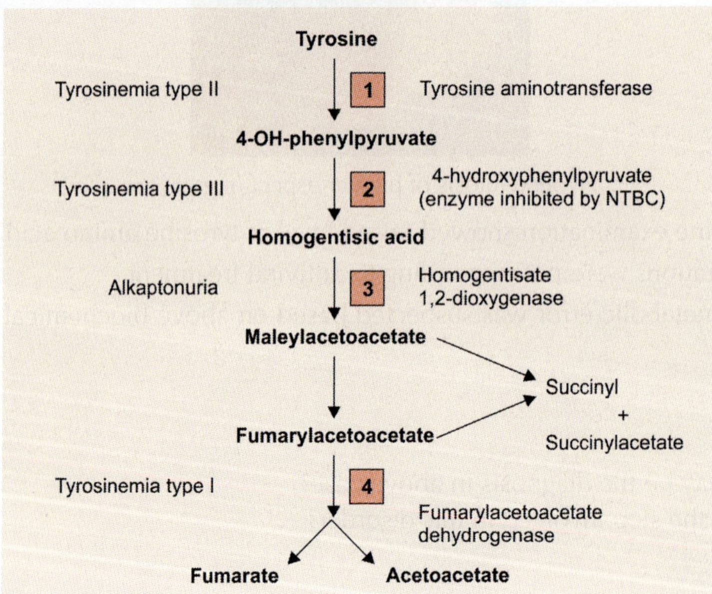

Fig. 2: Tyrosine catabolism and steps involved in various disorders along with respective enzyme deficiency are shown

Bibliography

1. Benatiya AI, Bouayed MA, Touiza E, Daoudi K, Bhalil S, Elmesbahi I, Tahri H. La tyrosinemie type II. A propos d'un cas (Tyrosinemia type II. Case report). Bull Soc BelgeOphtalmol. 2005;(296):57–61. French. PMID: 16050420.
2. Macsai MS, Schwartz TL, Hinkle D, Hummel MB, Mulhern MG, Rootman D. Tyrosinemia type II: nine cases of ocular signs and symptoms. Am J Ophthalmol. 2001 Oct;132(4):522–7. doi: 10.1016/s0002-9394(01)01160-6. PMID: 11589874.

CASE

35

Alkaptonuria

A 42-year-old male presents with complain of back pain and knee pain. His ear pinnae and sclera show black pigmentation. Skin of both the palms also shows blackish pigmentation. Urine turns black after few hours of exposure to atmospheric oxygen. X-ray spine shows narrowing of intervertebral disc and lumbar lordosis*.

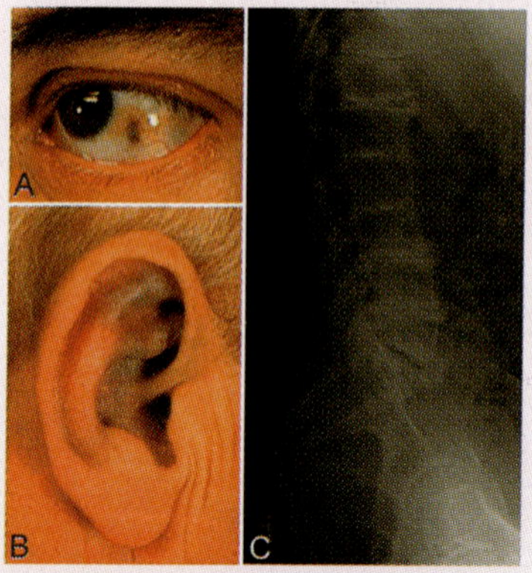

Black pigmentation sclera, ear pinnae. X-ray showing lumbar lordosis and narrow intervertebral disc

QUESTIONS

Q.1. Comment on the clinical condition this patient may be suffering from.

Q.2. What is the reason of black discoloration of various body tissues?

Q.3. How to suspect the diagnosis of alkaptonuria?

Q.4. What is the treatment suggested in this patient?

Explanations

Ans.1. This male has typical presentation of alkaptonuria.

Alkaptonuria also called **"endogenous ochronosis"** is an autosomal recessive disorder which is due to **deficiency of enzyme homogentisate 1,2-dioxygenase** also known as **homogentisic acid oxidase (HGA oxidase)**, involved in phenylalanine and tyrosine catabolism.

Gene for this enzyme is located on chromosome 3q2.

In deficiency of this enzyme, homogentisic acid (HGA) is not further metabolized.

Ans.2.

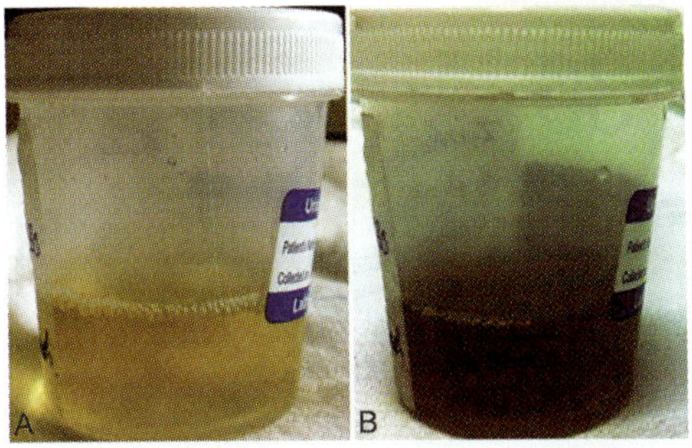

Fig. 1: Fresh urine and discoloration of urine on standing in alkaptonuria

- Homogentisic acid (HGA) is a colorless substance which is excreted in urine and undergoes auto-oxidation to corresponding **quinone (benzoquinone acetate).**
- These quinones undergo polymerization to blackish pigment. "**Alkapton**" which gets deposited in soft tissues and also undergoes excretion in urine.
- HGA gets deposited in articular cartilages of knee, hip and shoulder and resulting in arthritis of these joints. Intervertebral disc is also the site where such deposit results in back pain. Deposition of these pigments can be visualized in ear pinnae.
- **Deposition of these pigment in these tissues is termed "ochronosis" due to ochre color of the deposits.**
- The discoloration tends to be most pronounced on sun-exposed sites, cartilage of the ears and nose, areas of high eccrine sweat gland density, such as axillae, palms soles, and genitalia.
- In majority of such patients, pigment may get deposited in endocardium, aortic and coronary arteries resulting in cardiac involvement.

Ans.3.

- Fresh urine is colorless but on exposure to atmospheric oxygen urine turns black. Addition of alkali (10% ammonia) enhances black discoloration of urine.
- Black staining of diaper is telltale sign of possibility of presence of alkaptonuria in pediatric age group. Skin pigmentation is mostly late to appear and is noticed in third or fouth decade of life.
- Presentation is mostly late in alkaptonuria when patient may come with history of back pain, difficulty in walking and black discoloration of skin, eye, pinnae.
- Radiological investigation may reveal narrowing and calcification of intervertebral disc, fusion of vertebral bodies, lumbar lordosis.

Ans.4. Treatment with vitamin C, 500 mg/day twice daily along with protein restriction is helpful. Milk, meat, poultry, nut, cheese, egg to be avoided. NSAIDs* advised for arthropathy along with physical exercise.

Bibliography

1. Craide FH, Fonseca JS, Mariano PC, Fernandez NM, Castro CG, MeneYde S. Alkaptonuria--case report. An Bras Dermatol. 2014 Sep-Oct;89(5):799–801. doi: 10.1590/abd1806-4841.20143052. PMID: 25184921; PMCID: PMC4155960.
2. Damarla N, Linga P, Goyal M, Tadisina SR, Reddy GS, Bommisetti H. Alkaptonuria: A case report. Indian J Ophthalmol. 2017 Jun;65(6):518–21. doi: 10.4103/ijo.IJO_337_16. PMID: 28643719; PMCID: PMC5508465.

Word Meaning

Lumbar lordosis: Inward curve of the lumbar spine in lower back.

NSAIDs: Nonsteroidal anti-inflammatory drugs.

CASE

36

Hartnup's Disease

A 12-year-old girl presents with history of one episode of **generalized fits** and repeated episodes of **red scaly skin rashes** for past 2 years. She was born at full term with normal milestones* to a nonconsanguineous* parents. She also has history of psychotic behavior and difficulty in walking for past 6 months.

Urinary investigation by chromatography revealed presence of multiple neutral amino acid and presence of **indican in the urine.**

NMR spectroscopy* finding revealed excess excretion of following amino acids:

- Tryptophan
- Phenylalanine
- Tyrosine
- Leucine
- Isoleucine
- Valine
- Alanine
- Serine
- Threonine
- Asparagine

Diagnosis of Hartnup's disease is made.

QUESTIONS

Q.1. What is the etiopathogenesis of Hartnup's disease?

Q.2. What is the reason for skin rashes in this child?

Q.3. How to confirm the diagnosis?

Q.4. How to detect indican in the urine?

Explanations

Ans.1. Hartnup's disease is due to **mutation in neutral amino acid transporter** in kidney and intestine. Characteristic features of this disease are dermatological, neurological and psychiatric. Patient presents with pellagra like skin rash, ataxia and psychotic manifestation.

***SLC6A19* gene** located on chromosome **5p15.33** is responsible for encoding a **neutral amino acid transporter B(0) AT1,** which is responsible for transportation of various monoamino monocarboxylic amino acids (neutral amino acid) from lumen to intracellular location.

Mutation of this gene as a cause for Hartnup's disease is reported in many cases. Lack of normal transportation due to such mutations results in loss of various neutral amino acids in urine and stool of patients, resulting in severe **hypoaminoacidemia** which is responsible for various symptoms and signs of the disease.

Ans.2. Lack of tryptophan causes deficient synthesis of nicotinamide which is probably the reason behind such **pellagra-like manifestation** in Hartnup's disease and also responsible for neurological features. Role of tryptophan in synthesis of nicotinamide in kynurenine pathway is depicted in Fig. 2.

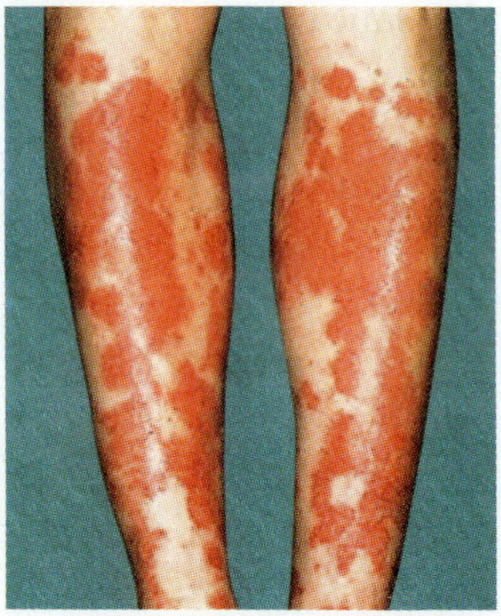

Fig. 1: Red scaly skin rash on lower limb

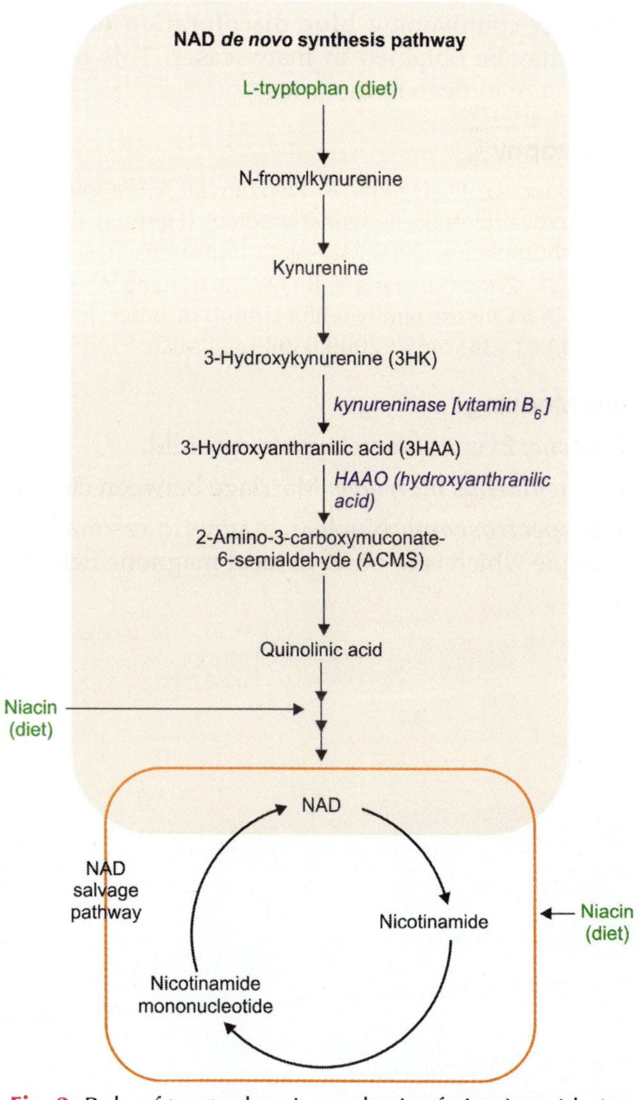

Fig. 2: Role of tryptophan in synthesis of nicotinamide (note the role of vitamin B_6 as cofactor for kynureninase enzyme)

Cerebellar ataxia is most characteristic neurological finding in patient with Hartnup's disease. Skin rash is seen on face, neck, arm and dorsum of feet and are more pronounced on exposure to sunlight.

Occasionally, acrodermatitis enteropathica like skin leisons in presence of normal zinc level has been reported in patient with Hartnup's disease.

Ans.3. Excretion of excess amount of neutral amino acid in the urine detected by chromatography and **NMR spectroscopy** along with characteristic clinical history and genetic analysis for mutation detection are used for diagnosis.

Ans.4. Unabsorbed tryptophan in intestinal lumen is acted upon by intestinal bacteria which convert it to **indole compounds** which is absorbed in general circulation only to get excreted in urine. **Indican in the urine** can be detected by **Obermeyer test.**

Mother complaining **blue discoloration of baby diaper** is characteristic history which may be obtained in many cases. This blue discoloration of diaper is due to excretion of indican in the urine.

Bibliography

1. Milovanoviæ D, Djukiæ A, Stepanoviæ R, Pekoviæ D, Vranjeseviæ D. Hartnupovabolest (prikazdvabolesnika is jedneporodice) [Hartnup disease (report of 2 cases in one family)]. SrpArhCelok Lek. 2000 Mar-Apr;128(3-4):97–103. Serbian. PMID: 10932618.
2. Zheng Y, Zhou C, Huang Y, Bu D, Zhu X, Jiang W. A novel missense mutation in the SLC6A19 gene in a Chinese family with Hartnup disorder. Int J Dermatol. 2009 Apr;48(4):388–92. doi: 10.1111/j.1365?4632.2009.03989.x. PMID: 19335424.

Word Meaning

Milestone: Event of development of child.

Consanguineous marriage: Marriage between close relatives.

NMR spectroscopy: Nuclear magnetic resonance spectroscopy. A spectroscopic technique which is to observe local magnetic field around atomic nuclei.

CASE

37

B_6 Pellagra

A 37-year-old male comes to OPD with complain of **numbness and tingling sensation** on fingers of both hands for past 2 weeks. He is on treatment for tuberculosis and is taking **isoniazid in antitubercular treatment (ATT) regime.**

Blood investigation showed low level of pyridoxal phosphate as 3 ng/ml (normal is >5 ng/ml). **Xanthurenic acid** is detected in urine.

QUESTIONS

Q.1. Which vitamin deficiency may be responsible for above manifestation in this patient? How the treatment with isoniazid is related with deficiency of this vitamin?

Q.2. Disturbance of which metabolic pathway results in excretion of xanthurenic acid in the urine and why?

Q.3. What is the treatment of this patient?

Q.4. Name other conditions where pellegra may be an associated finding.

Explanations

Ans.1. This patient is exhibiting the symptoms of deficiency of vitamin B_6. Deficiency of vitamin B_6 manifests as pellagra-like syndrome, with **seborrheic dermatitis*, glossitis*, and cheilosis***, and, in adults, can cause **depression, confusion, electroencephalogram (EEG) abnormalities, and seizures.**

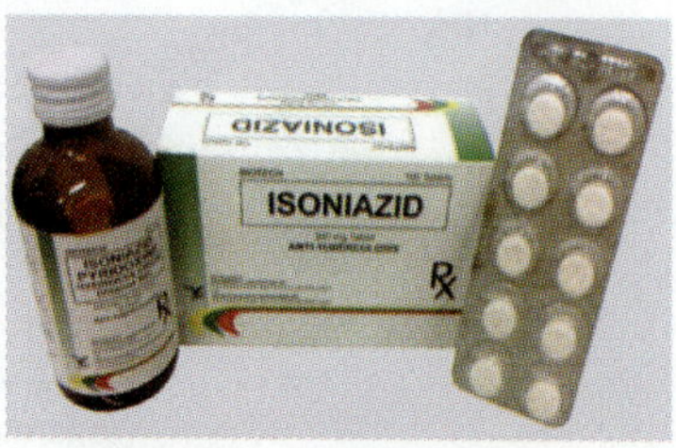

Fig. 1: Isoniazid (INH) used as ATT (antitubercular drug)

Therapy with **isoniazid (INH)** causes deficiency of vitamin B_6 (PLP). This is due to inhibition of **pyridoxine phosphokinase** by INH. This enzyme is necessary to activate pyridoxine to pyridoxal 5′ phosphate (PLP).

Another mechanism explained for such deficiency in INH treatment is direct interaction of INH with pyridoxine to make **hydrazone complex** which inactivate vitamin B_6.

Other causes of vitamin B_6 deficiency are as follows:

1. Dietary (rare)
2. Alcohol intake (common)
3. Protein energy malnutrition
4. Malabsorption
5. Hemodialysis
6. *Drugs:* Isoniazid (INH), cycloserine, hydralazine, corticosteroids, penicillamine, oral contraceptives.

Lack of PLP thus is responsible for peripheral neuropathy and other features of vitamin B_6 deficiency.

Ans.2. Disturbance of **kynurenine pathway** of tryptophan metabolism is seen in vitamin B_6 deficiency. This pathway synthesizes niacin (nicotinic acid) and needs PLP as cofactor for **kynureninase** enzyme.

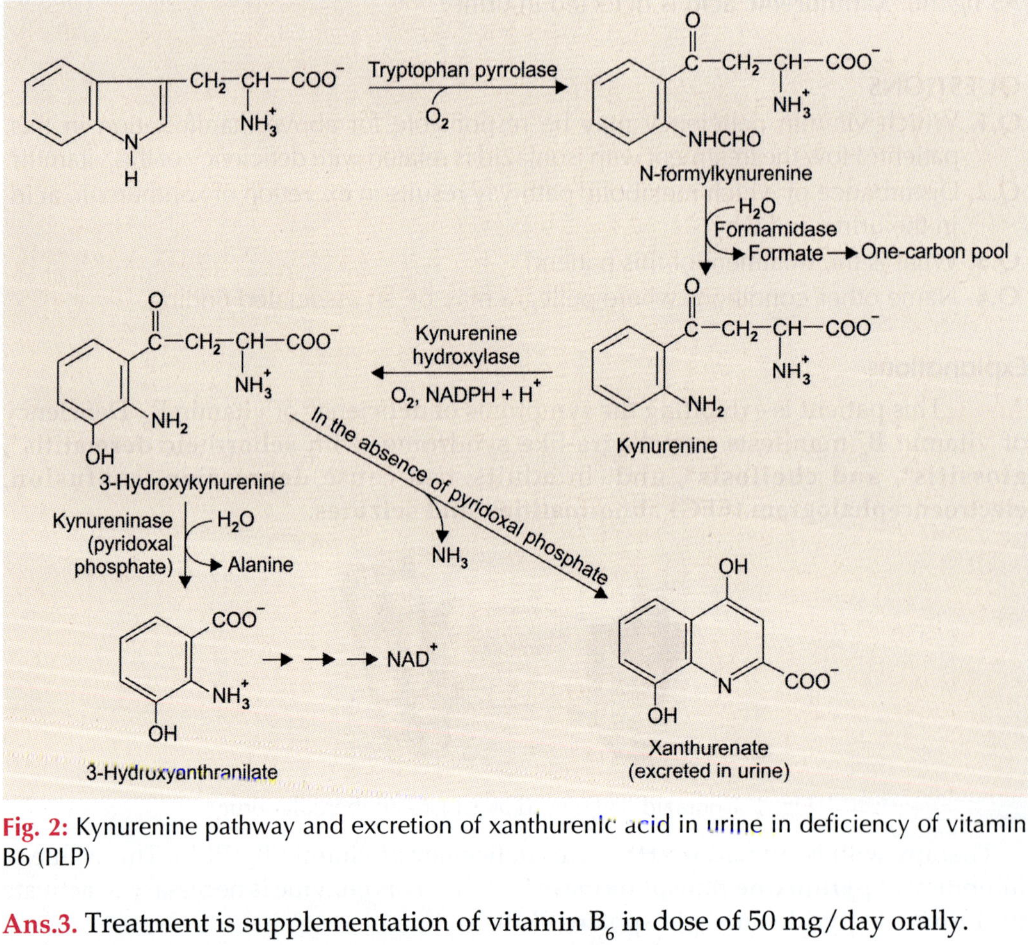

Fig. 2: Kynurenine pathway and excretion of xanthurenic acid in urine in deficiency of vitamin B6 (PLP)

Ans.3. Treatment is supplementation of vitamin B_6 in dose of 50 mg/day orally.

Ans.4. Other conditions which may be associated with pellegra are

a. Vitamin B_3 deficiency
b. Hartnup's disease
c. Carcinoid syndrome
d. Excess consumption of jowar
e. Maize consumed as staple diet.

Bibliography

1. Kipsang JK, Choge JK, Marinda PA, Khayeka-Wandabwa C. Pellagra in isoniazid preventive and antiretroviral therapy. IDCases. 2019 May 3;17:e00550. doi: 10.1016/j.idcr.2019.e00550. PMID: 31193074; PMCID: PMC6515148.
2. Segula D, Banda P, Mulambia C, Kumwenda JJ. Case report: A forgotten dermatological disease. Malawi Med J. 2012 Mar;24(1):19–20. PMID: 23638264; PMCID: PMC3588195.

Word Meaning

Seborrheic dermatitis: Red, scaly, itchy, greasy skin.

Cheilosis: Cracks and pain at corner of mouth.

Glossitis: Inflammation of tongue.

CASE

38

Carcinoid Syndrome

A 52-year-old male presents with case of repeated episodes of redness of upper body specially neck and face which in many circumstances is associated with diarrhea. Patient feel warmth and itchiness. He has noticed that these episodes are precipitated when he exercises or consume cheese. These episodes typically last for 15 to 20 minutes.

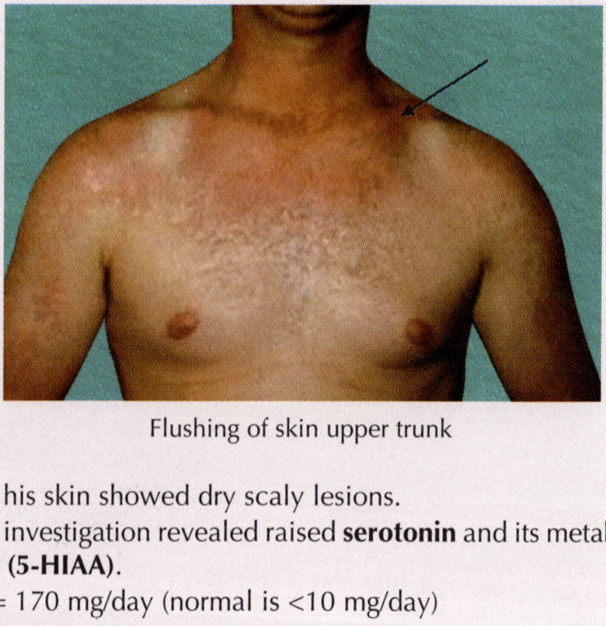

Flushing of skin upper trunk

On examination, his skin showed dry scaly lesions.

Plasma and urine investigation revealed raised **serotonin** and its metabolite 5-hydroxy-indoleacetic acid **(5-HIAA)**.

Urinary 5-HIAA = 170 mg/day (normal is <10 mg/day)

QUESTIONS

Q.1. What may be the diagnosis in this case?

Q.2. What may be the biochemical explanation of clinical signs and symptoms (diarrhea, flushing, dermatitis)

Q.3. What dietary advise is to be given to patients before urine collection for 5-HIAA measurement?

Q.4. What is the treatment advised in such cases?

Explanations

Ans.1. Typical clinical presentation of **flushing associated with diarrhea** along with biochemical finding of raised level of serotonin and its metabolite **5-hydroxyindole-acetic acid (5-HIAA)** strongly suggests the possibility of **carcinoid syndrome** in this patient. Carcinoid syndrome is tumor of **neuroendocrine cells** in the gastrointestinal tract which secrete excessive amount of serotonin (Fig. 1).

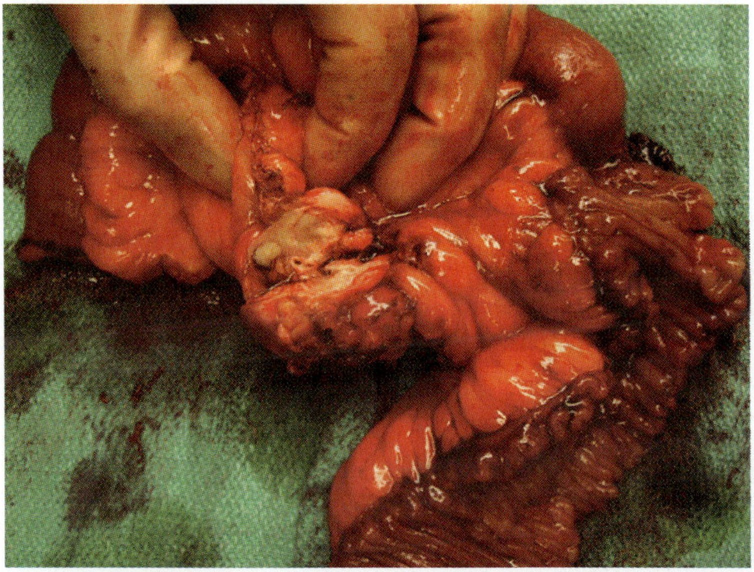

Fig. 1: Carcinoid tumor of small intestine

Ans.2. **Biochemical basis of signs and symptoms.**

Diarrhea: Serotonin has multiple biological effects like increased intestinal secretion and at the same time reduced absorption. This along with its impact of increasing intestinal motility results in diarrhea.

Flushing: It is also due to excessive secretion of serotonin.

Dermatitis: Serotonin is synthesized from tryptophan and in this condition almost half of the dietary tryptophan is diverted for making of serotonin, depriving the synthesis of niacin from tryptophan. Lack of niacin is the reason of pellagra like dermatitis in a case of carcinoid syndrome (Fig. 2).

Ans.3. **Serotonin-rich foods like banana, walnut, pineapple, avocados and medications like salicylates, cough syrup, acetaminophen** are to be avoided for 48 hours before starting the urine collection as these compounds give falsely high level of 5-HIAA.

Ans.4. Treatment is advised with **nicotinamide supplementation, diarrhea treated with loperamide, wheezing is treated with bronchodilators.**

Somatostatin analogue (octreotide) or serotonin receptor antagonists are advised.

Surgery is the curative treatment, if the tumor is localized.

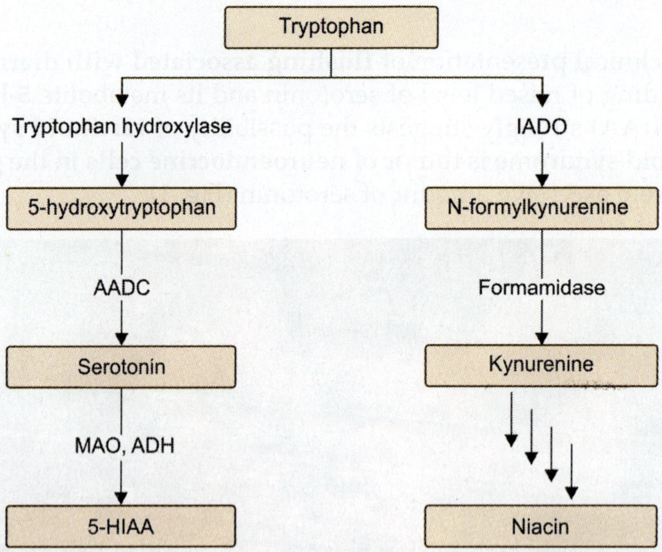

Fig. 2: Diversion of tryptophan for serotonin overproduction in carcinoid syndrome

Bibliography

1. Mahmood, Tallat, MD; Mudad, Raja, MD. Carcinoid Syndrome with Unknown Primary, American Journal of Clinical Oncology: August 1999;22(4):384–86.
2. Naalla R, Konchada K, Kannappan O, Lingadakai R. Duodenal carcinoid with carcinoid syndrome. BMJ Case Rep. 2014 Jan 10;2014:bcr2013202159. doi: 10.1136/bcr-2013-202159. PMID: 24414187; PMCID: PMC3902501.

CASE

39

Classical or Typical Homocystinuria

An 8-year-old female presents with complain of involuntary repeated movements in left hand for past 4 months, which has been progressively worsening over the period of one month. She also complains of difficulty in vision for past 2 years. Baby is born to consanguineous marriage. Her elder brother died at the age of 4 years after few episodes of fits.

Blood investigations revealed high level of **serum homocysteine 220 μmol/L (normal is <15 μmol/L).**

LFT, KFT, TFT, serum electrolyte results were within normal limits.

Patient was treated with high dose of pyridoxine (vitamin B_6) and methionine restriction was done. Patient dystonic movements showed improvement after 4 months of such treatment.

QUESTIONS

Q.1. What may the disorder this child may be suffering from?

Q.2. How the homocysteine is metabolized biochemically?

Q.3. What is the rationale behind giving high dose of pyridoxine?

Explanations

Ans.1. This child may be suffering with **classical HCU.** It is **autosomal recessive** disorder due to deficiency of enzyme **cystathionine beta synthase.**

This disease involves multiple organ system, four-organ system predominantly affected are **ocular system, nervous system, vascular system and skeletal system.**

Clinical manifestations involve eye (subluxation of lens*, secondary glaucoma*, optic atrophy*, retinal detachment), central nervous system (cerebellar ataxia, nystagmus*, aphasia*, hypotonia and muscle weakness), vascular (thromboembolic episodes and atherosclerosis), and **skeleton (arachnodactyly*, lengthening of long bones, osteoporosis of vertebrae and long bones, scoliosis*).**

Mild to moderate degree of mental retardation is seen in 50 to 75% of patients. Seizures, dystonia, behavior and personality disorders are additional findings.

Dystonia* may be because of microinfarct in brain or may be due to direct toxic effect of homocysteine.

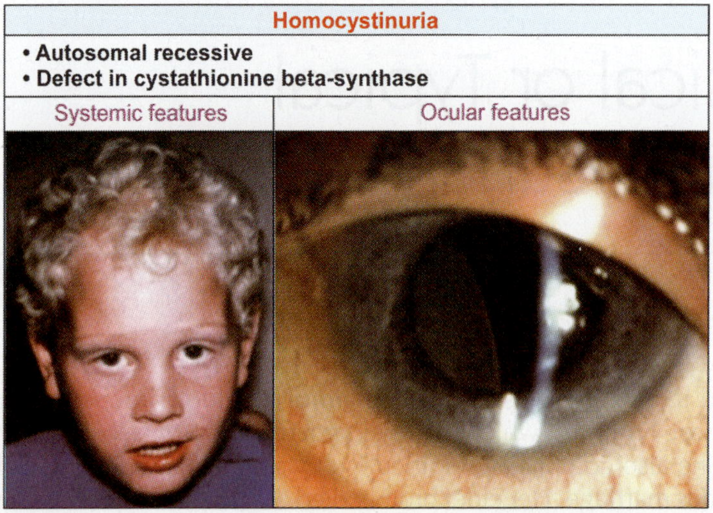

Fig. 1: Clinical presentation of HCU. Note: Fine wooly hair, subluxation lens

Thromboembolic episodes are seen due to vascular endothelial dysfunction and lipid peroxidation. Cerebrovascular accident, pulmonary embolism, carotid, and renal artery thrombosis and lower gastrointestinal bleed have been reported as complications of homocystinuria. HCU may be associated with **aortic aneurysm.**

Skin manifestations are **malar flush**, **cutis reticulata**, thin hair.

When exposed to low temperature skin develops characteristic red and blue articulated vascular pattern known as **cutis marmorata.**

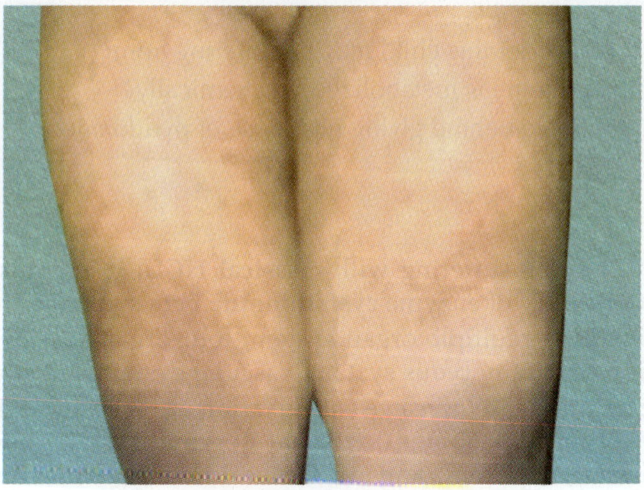

Fig. 2: Cutis marmorata

Ans.2. Homocysteine is intermediate of methionine metabolism which may either undergo **trans-sulfuration reaction** or may undergo **transmethylation.**

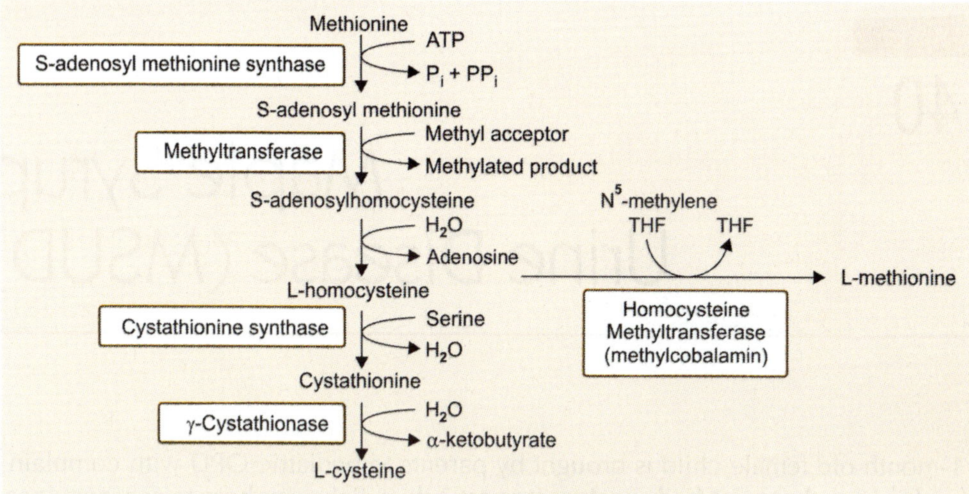

Fig. 3: Methionine metabolism and fate of homocysteine

Ans.3. Vitamin B_6 is cofactor for action of **cystathionine beta synthase enzyme. High dose of vitamin B_6 supplementation** may help as it may increase residual activity of cystathionine beta synthase. Supplementation of **betain** helps in remethylation of homocysteine and lowers its level. In addition, folic acid and vitamin B_{12} can be supplemented. This along with restriction of methionine intake is proved to be useful in such cases. **Target of therapy is to maintain the level of homocysteine <15 µmol/L.**

HCU is second most common treatable aminoacidoapthy after phenylketonuria with the worldwide incidence of 1 in 50,000 live births.

HCU may resemble Marfan syndrome. It can be differentiated by **cyanide nitroprusside test** in urine which is positive for homocysteine, hence positive in a case of homocystinuria.

Bibliography

1. Li SC, Stewart PM. Homocystinuria and psychiatric disorder: a case report. Pathology. 1999 Aug;31(3):221–4. doi: 10.1080/00313029910502 5. PMID: 10503267.
2. Rezazadeh A, Oveisgharan S, Shahidi G, Naghdi R. A Case Report of Homocystinuria With Dystonia and Stroke. Child Neurol Open. 2014 Aug 26;1(1):2329048X14545870. doi: 10.1177/2329048X14545870. PMID: 28503581; PMCID: PMC5417037.
3. Saini AG, Padmanabha H, Attri S, *et al*. Malar rash in classical homocystinuria. *Case Reports* 2017;**2017:**bcr-2017-220296.

Word Meaning

Subluxation of lens: Displacement or dislocation of lens from its natural position.
Glaucoma: High pressure in eye which damages optic nerve.
Optic atrophy: Wasting of optic nerve.
Arachnodactyly: Long, thin, slender fingers.
Scoliosis: Lateral curvature of spine.
Dystonia: Muscle spasm and abnormal posture.

CASE

40

Maple Syrup Urine Disease (MSUD)

A 1-month-old female child is brought by parents to pediatric OPD with complain of refusal to feed, **dystonic* body posture** for past 2 days. Baby was born to **consanguineous** parents and was apparently normal at birth. Her condition **progressively deteriorated** over past two days. Her elder brother died at the age of two years due to unknown disease.

Blood and urine investigations were done.

Urine of the baby gave characteristic **fruity/burnt sugar odor.**

Plasma and urine showed elevated level of **valine, leucine and isoleucine** in tandem mass spectrometry.

QUESTIONS

Q.1. Comment on the clinical condition baby is suffering from.

Q.2. What strategy to adopted for early diagnosis in such cases?

Q.3. What is the treatment advised?

Explanations

Ans.1. This baby is suffering with **maple syrup urine disease (MSUD).** MSUD is also known as **branched chain ketoaciduria (oxoaciduria).** It is an **autosomal recessive disorder** which is due to deficiency of enzyme. **"Branched-chain α-keto acid dehydrogenase complex (BCKD)"**. This enzyme complex is required at second step during branched-chain amino acid catabolism (Fig. 1).

This enzyme complex helps in the **oxidative decarboxylation** of branched chain α-keto acids of branched-chain amino acid (BCA) to their respective branched chain acetyl-CoA.

This enzyme complex has three enzymes: E1, E2, E3. Mutation may be on any of the gene encoding enzymes E1, E2 or E3 of the enzyme complex and this disease presents with different severities depending upon the extent of deficiency of these enzymes.

MSUD is classified into five clinical phenotypes:

i. Classic
ii. Intermediate

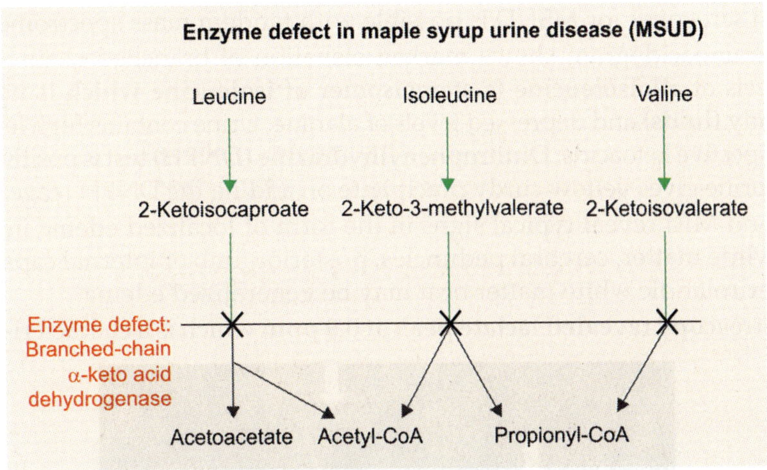

Fig. 1: The step involved in MSUD

iii. Intermittent
iv. Thiamine responsive
v. **Dihydrolipoyl dehydrogenase** deficient form

Classic type is most common and **most severe type** which presents in newborn period as poor feeding, dystonia, vomiting and seizures. Children are normal at birth but neurological deterioration is observed in first 3 to 4 days of birth when they present with **apnea, seizures and death**. If they survive, severe **mental retardation** is inevitable.

Accumulation of leucine is predominantly responsible for neurological symptoms. The elevated isoleucine level causes the **peculiar maple syrup odor (boiled Chinese herb)** of urine.

Ans.2. Prenatal diagnosis can be performed by either by measuring the enzyme activity in cultured **amniocytes or chorionic villous cells** or by mutation analysis. The level of branched-chain amino acid (BCA) may be measured in amniotic fluid which when raised gives fair idea regarding this disease.

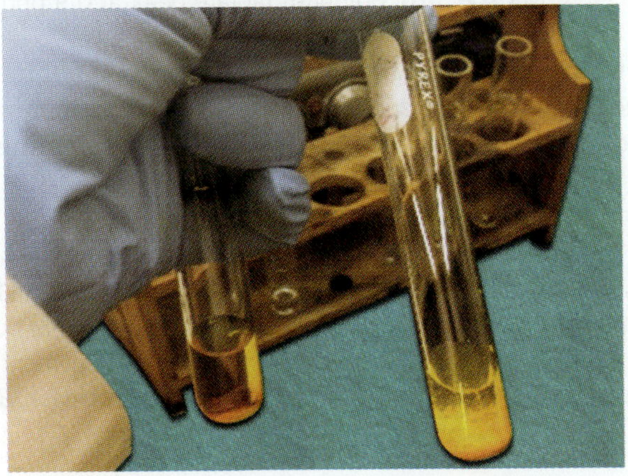

Fig. 2: Positive dinitrophenylhydrazine (DNPH) test in a case of MSUD (maple syrup urine disease)

Newborn screening for MSUD is possible with tandem mass spectrometry (TMS).

Plasma aminoacidogram shows marked elevation of branched chain amino acid, elevated levels of **alloisoleucine (a stereoisomer of isoleucine which is not normally found in body fluids)** and depressed levels of alanine. Urine contains high levels of BCA and their respective ketoacids. **Dinitrophenylhydrazine (DNPH) test** is positive in MSUD. In this test, urine gives **yellow curdy precipitate** on adding the DNPH reagent (Fig. 2).

The CT and MRI reveal typical signs in the form of localized edema in brainstem, cerebellar white matter, cerebral peduncles, posterior limb of internal capsule, globus pallidus, perirolandic white matter or it may be generalized edema.

MR spectroscopy revealed lactate peak at 0.9 ppm which is a characteristic finding.

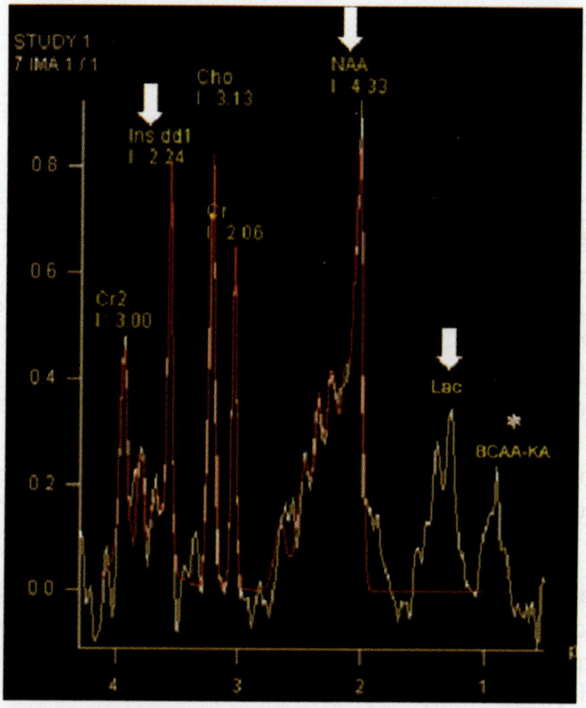

Fig. 3: MR spectroscopy with lactate peak at 0.9 ppm

Ans.3. Treatment of MSUD is formula diet where restriction of branched-chain amino acid specially leucine is done. This is accompanied with supplementation of high doses of thiamine (75 mg thrice daily) for three weeks. This is useful in **thiamine responsive MSUD**.

Bibliography

1. Yeung CY, Lin SP, Kao HA, Shen EY, Lee HC, Lin CH, Hsiao KJ. Maple syrup urine disease: report of two cases. J Formos Med Assoc. 1993 Aug;92(8):765–8. PMID: 7904856.
2. araspongUaariyapanichkul, PuthitaSaengpanit, PonghataiDamrongphol, Kanya Suphapeetiporn, SirinuchChomtho, "Skin Lesions Associated with Nutritional Management of Maple Syrup Urine Disease", *Case Reports in Dermatological Medicine*, vol. 2017, Article ID 3905658, 4 pages, 2017.

Word Meaning

Dystonic body posture: Abnormal body posture due to muscle spasm.

CASE

41

Isovaleric Acidemia

A 4-month-old female baby is brought with history of fever and cough for past 1 week. She has difficulty in breathing and has **Kussmaul respiration*.** She is lethargic* and dehydrated. Mother complains of strange odor in the urine.

Plasma amino acids and urine for organic acids were sent to evaluate for underlying metabolic disorders, if any. Urine had **cheesy odor.**

Urine organic acid analysis by gas chromatography–mass spectrometry showed marked excretion of **3-hydroxybutyric acid** along with moderate excretion of **3-hydroxy-isovaleric acid.**

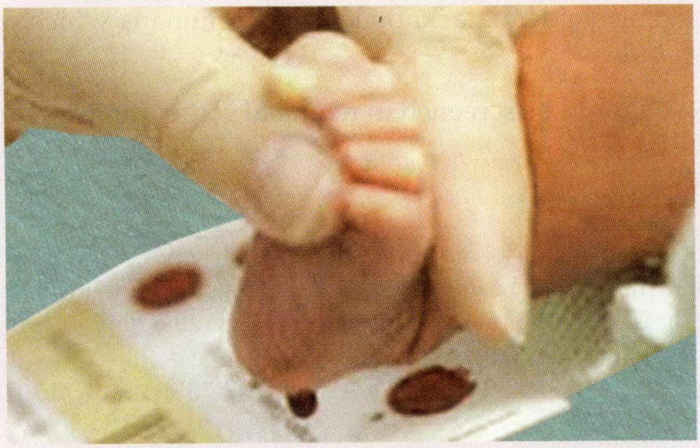

Heel prick for neonatal screening

QUESTIONS

Q.1. What is the diagnosis in this child?

Q.2. Which pathway is affected?

Q.3. What is the treatment plan?

Explanations

Ans.1. Raised level of **3-hydroxybutyric acid** along with **3-hydroxyisovaleric acid** denotes that this patient is having **isovaleric acidemia (IVA).** This is an **autosomal**

recessive disorder which is due to deficiency of enzyme **IVD (isovaleryl-CoA dehydrogenase.** This disease is precipitated in this child due to respiratory infection.

Ans.2. The enzyme IVD **(isovaleryl-CoA dehydrogenase)** is required in leucine catabolism. Deficiency o this enzyme results in isovaleryl-CoA accumulation in plasma and formation of various organic acids which are excreted in urine. Urine and other body fluids have characteristic ***cheesy odor or sweaty feet odor.***

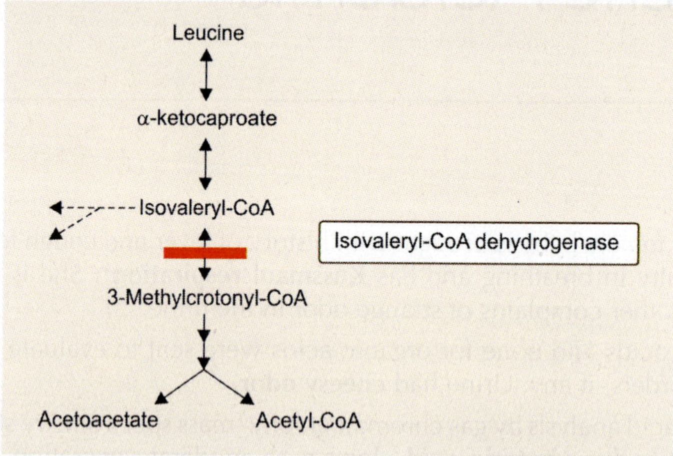

Fig. 1: Leucine catabolism and step blocked in isovaleric acidemia

It is earliest organic acidemia acknowledged in humans which results in serious mortality and morbidity.

Ans.3. This disease should be screened in **neonatal screening programme.** Treatment is supportive care.

Bibliography

1. Haque N, Saha N, Alam MB, Mohammad Q, Hoque S, Islam A, Chowdhury Y, and Debnath, B. Isovaleric Acidemia in a 5 Years of Boy: A Case Report. Journal of Current and Advance Medical Research, 2019;6(1):64–66. https://doi.org/10.3329/jcamr.v6i1.40788
2. Khan A, Zahid B, Khan S, Ahmad SA. Isovaleric Acidemia: A Rare Case of an Inborn Error of Metabolism. Cureus. 2020 Feb 29;12(2):e7150. doi: 10.7759/cureus.7150. PMID: 32257695; PMCID: PMC7108672.
3. Kiliç M, Kaymaz N, Özgül RK. Isovaleric acidemia presenting as diabetic ketoacidosis: a case report. J Clin Res Pediatr Endocrinol. 2014;6(1):59–61. doi: 10.4274/Jcrpe.1181. PMID: 24637313; PMCID: PMC3986742.

Word Meaning

Kussmaul respiration: Deep and rapid respiration.

Lethargy: Loss of energy and ethusiasm.

CASE
42

Myocardial Infarction and Enzyme Marker

A 56-year-old chronic smoker for past 30 years who is a lawyer by profession presents with heaviness in chest with pain on left upper limb for past 12 hours. He is known hypertensive on regular antihypertensive medication. He is not a known diabetic.

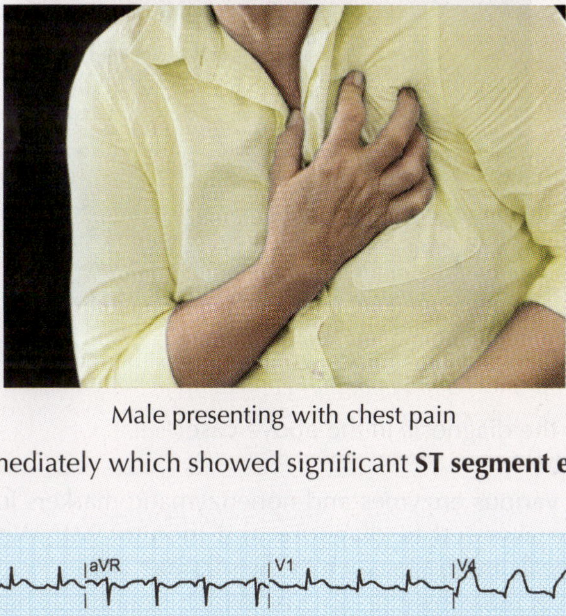

Male presenting with chest pain

ECG was done immediately which showed significant **ST segment elevation** in I, II and V1 to V6 leads

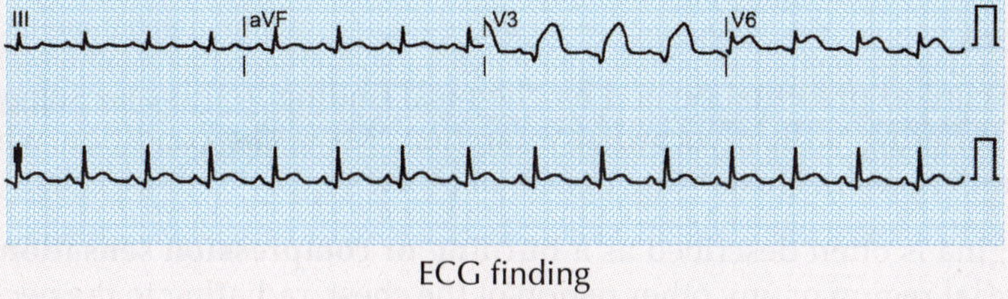

ECG finding

Following are the results of various laboratory investigations:

Troponin T = 134 ng/ml (normal = <5 ng/ml)

Troponin I = 240 ng/ml (normal <0.04 ng/ml)

CPK-Total = 2000 IU/L (normal = 40–300 IU/L)

CPK-MB = 340 IU/L (normal 5–25 IU/L)

AST = 765 IU/L (normal = 15–45 IU/L)

LDH = 323 IU/L (normal = 140–280 IU/L)

Angiography revealed **95% block** in proximal portion of **left anterior descending (LAD)** artery.

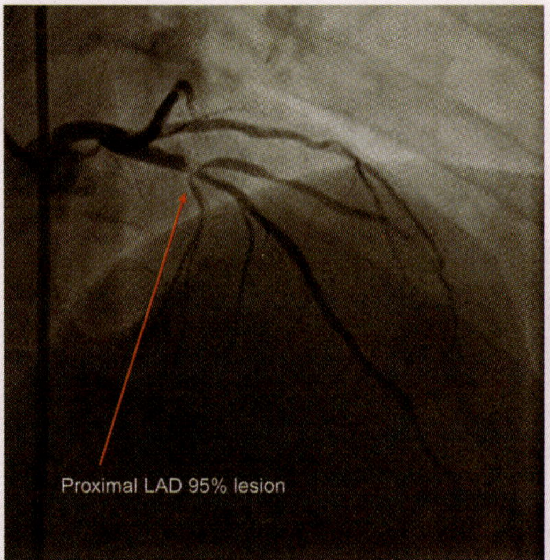

QUESTIONS

Q.1. What may be the diagnosis in the above case?

Q.2. What is angina?

Q.3. What are the various enzymes and nonenzymatic markers for assessment of MI which can be assessed to diagnose and monitor MI? What is time of their appearance, peak and persistence in the plasma?

Q.4. Why troponin is more specific for MI? Which troponin is better for detection of MI: Troponin T or Troponin I?

Q.5. Name the enzyme/s which may be used for treatment in such cases.

Q.6. What are the risk factors for myocardial infarction?

Explanations

Ans.1. The clinical presentation and biochemical finding in this case suggest that this patient is suffering with myocardial infarction. MI is the condition when cardiac myocyte **undergoes necrosis*** due to lack of blood supply to those cells.

Ans.2. **Angina** is often described as a **burning or compression sensation located in the precordial region** or any other region of the chest, radiating to the neck, shoulder and left arm. It usually increases in intensity within minutes and may be accompanied by symptoms such as nausea and sweating. It can be triggered by physical or emotional stress and relieved by rest or use of nitrates.

Ans.3. There are various enzymes which are used in assessment of MI. They are:

CPK-total, CPK-MB, AST, LDH

Out of these, **AST and LDH** are nonspecific markers of MI as these enzymes are found in other cells also in addition to heart. They have a wide tissue distribution that significantly limits the specificity for myocardial necrosis.

CPK is enzyme which serves as carrier of high energy phosphate and has got three isoenzymes, namely CPK-BB, CPK-MB, CCPK-MM. Creatine kinase MB (CK-MB) is an isoenzyme of creatine kinase that is most abundant in the heart and should be assessed to rule in or rule out the myocardial infarction.

Nonenzymatic markers of MI are troponin T and troponin I and myoglobin.

Below mentioned graph is well representing the time of onset, peak and duration of persistence of various enzymes in a typical case of MI.

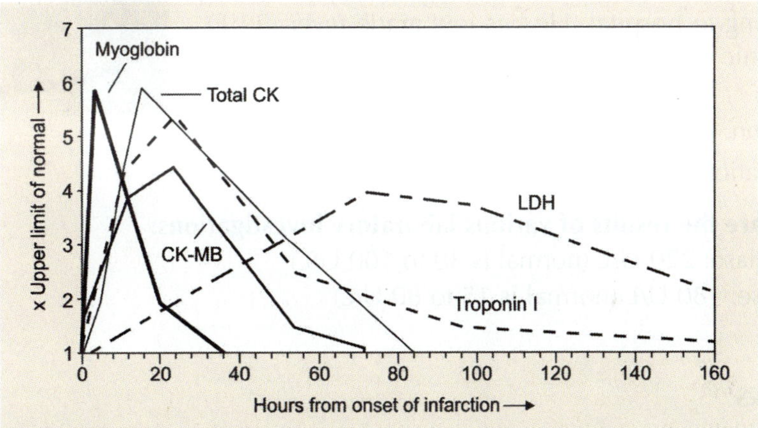

Fig. 1: Graphical representation of various enzymatic and nonenzymatic markers in case of MI

Ans.4. Troponin I is better than troponin T.

Ans.5. There are few enzymes which are being used as **thrombolytic agents** in such cases of MI. They are **streptokinase, urokinase,** etc.

Ans.6. Important risk factors for myocardial infarction are: Dyslipidemia, male gender, diabetes mellitus, hypertension, smoking, sedentary lifestyle, advanced age.

Bibliography

1. https://www.emra.org/emresident/article/acute-mi-case-report/
2. Ribeiro WN, Yamada AT, Benvenuti LA. Case 4/2014—A 66-year-old man with acute myocardial infarction and death in asystole after primary coronary angioplasty. Arq Bras Cardiol. 2014 Sep;103(3):e31–6. doi: 10.5935/abc.20140129. PMID: 25317943; PMCID: PMC4193075.

Word Meaning

Necrosis: Cell death.

CASE

43

Acute Pancreatitis

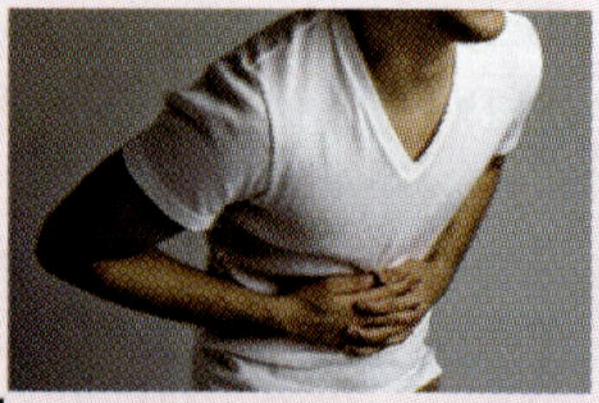

A 45-year-old male presents with intense pain in epigastric and periumbilical region. Pain is radiating to back and is associated with vomiting. He had two episodes of vomiting while coming to hospital. He has low grade fever (99°F). He is chronic alcoholic indulged in alcoholism for past 15 years.

On palpation, abdomen was found to be distended and tender.

On auscultation, bowel sounds were absent.

Following are the results of various laboratory investigations:

Serum amylase: 220 U/L (normal is 30 to 100 U/L)

Serum lipase: 180 U/L (normal is 15 to 80 U/L)

QUESTIONS

Q.1. What is the diagnosis based on clinical and biochemical findings?

Q.2. What is the criteria of diagnosis of acute pancreatitis?

Q.3. What is the single best enzyme to diagnose pancreatitis?

Q.4. What is the most sensitive test to detect pancreatitis in early stages?

Q.5. What are the common causes which are associated with acute pancreatitis?

Q.6. Which type of hyperlipidemia is associated with chances of getting acute pancreatitis?

Explanations

Ans.1. This patient is having **acute pancreatitis** as diagnosed based on clinical and biochemical findings.

Ans.2. Two of the following three criteria to be fulfilled for diagnosis of acute pancreatitis:

a. Epigastric pain radiating to back

b. Threefold or greater increase of serum lipase and/or amylase

c. Abdominal imaging studies giving the conclusive findings.

Ans.3. **Single best enzyme** to measure in acute pancreatitis is **lipase** which rises parallel to amylase but tends to stay longer in plasma than amylase.

Time duration of elevation and persistence of pancreatic enzyme in acute pancreatitis:

- Onset of enzyme elevation: within 24 hours
- Remain elevated: 3 to 7 days
- Return to normal: 7 days

Lipase is specific for pancreas and amylase is not specific.

Raised amylase may be seen in the following conditions:

- Acute pancreatitis
- Chronic pancreatitis (ductal obstruction)
- Pancreatic pseudocyst
- Pancreatic trauma
- Pancreatic cancer
- Renal insufficiency
- Cancer lung, ovary, breast esophagus (tumor hyperamylasemia)
- Macroamylasemia
- Burns
- DKA (diabetic ketoacidosis)
- Pregnancy
- Opiates

Ans.4. Direct invasive test to **detect pancreatic secretory function (secretin test)** is the most sensitive and specific test to detect early chronic pancreatic disease when imaging is inconclusive or normal.

Following imaging tests are done in a case of pancreatitis:

- **Non-invasive test:** USG, CT, MRCP (magnetic resonance cholangiopancreatography)
- **Invasive test: ERCP (endoscopic retrograde cholangiopancreatography)**, EUS (endoscopic ultrasonography)

Ans.5. ***Causes of acute pancreatitis***

- Gallstone
- Alcohol: Alcoholism and cigarette smoking compound the effect on incidence of pancreatitis.

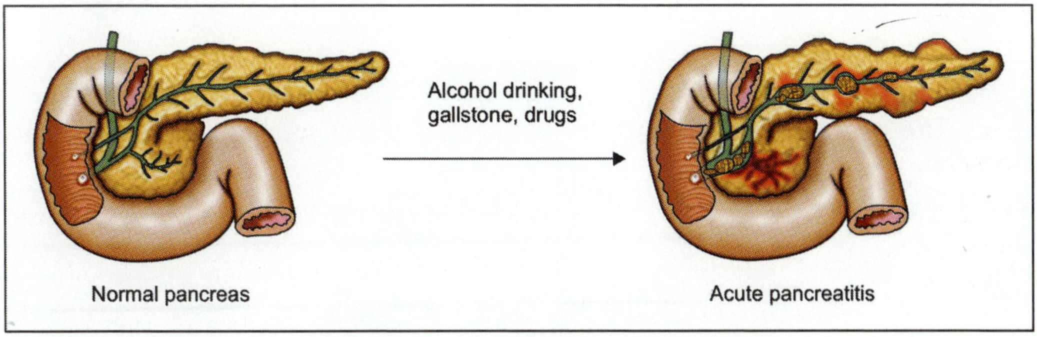

Fig. 1: Factors precipitating the acute pancreatitis

- Hypertriacylglyceridemia (>1000 mg/dl)
- Abdominal trauma
- ERCP induced

Ans.6. **Hypertriacylglyceridemia** either because of deficiency of lipoprotein lipase enzyme or because of Apo CII results in raised chylomicron in blood. This causes hypertriacylglyceridemia in lipid profile. This is associated with pancreatitis. Reason of association of hypertriglyceridemia and incidences of pancreatitis is obscure.

Bibliography

1. Aboulhosn K, Arnason T. Acute pancreatitis and severe hypertriglyceridaemia masking unsuspected underlying diabetic ketoacidosis. BMJ Case Rep. 2013 Sep 4;2013:bcr2013200431. doi: 10.1136/bcr-2013-200431. PMID: 24005972; PMCID: PMC3794188.
2. Nadhem O, Salh O: Acute Pancreatitis: An Atypical Presentation. Case Rep Gastroenterol 2017;11:359–63. doi: 10.1159/000475920

CASE

44

Ethanol used as Therapeutic Agent in Methanol Poisoning

A 29-year-old male comes to emergency unit with difficulty in vision and gasping respiration. His friend gives the history of alcohol intake from a local shop. Blood investigation revealed **high anion gap metabolic acidosis**.

HCO_3^-: 4.3 mmol/L (normal = 12 mmol/L)

pH: 6.89 (normal = 7.35–7.45)

Serum osmolality was 379 mOsm/kg

Na^+: 146 mmol/L (normal = 135–145 mmol/L)

K^+: 7.7 mmol/L (normal = 3.5–5 mmol/L)

Urea: 32 mg/dl (normal = 15–45 mg/dl)

Glucose: 108 mg/dl (normal random = 80–140 mg/dl)

Cl^-: 116 mmol/L (normal = 90–110 mmol/L)

Lactate: 14.2 mmol/L (normal = 0.5–1.0 mmol/L)

Patient was given **IV ethanol** for treatment in addition to dialysis and other supportive care.

QUESTIONS

Q.1. What is the rationale behind giving ethanol in methanol poisoning?

Q.2. Explain the occurrence of lactic acidosis in methanol poisoning.

Explanations

Ans.1. Methanol is a toxic alcohol. Toxicity of methanol is due to formation of toxic compounds **formaldehyde** during its metabolism. Formaldehyde is a toxic gas which damages the retina which may result in **blindness.**

Ethanol acts as a **competitive inhibitor of alcohol dehydrogenase enzyme.** In presence of ethanol, enzyme **alcohol dehydrogenase preferentially binds ethanol which is converted to acetaldehyde. Acetaldehyde thus produced is not** of immediate toxicity, hence infusion of ethanol provides protection against toxicity of formaldehyde. Goal of therapy is to maintain **serum level of ethanl as 80 to 120 mg/dl.**

Hemodialysis is also useful in treatment of **methanol poisoning** to get rid of toxic compounds already formed.

Ans.2. Methanol metabolism produces excess amount of NADH. This is due to the fact that **alcohol dehydrogenase** requires NAD as coenzyme which produces NADH. Further metabolism of formaldehyde needs **aldehyde dehydrogenase** which also requires NAD as cofactor producing NADH. Figure 1 explains this process:

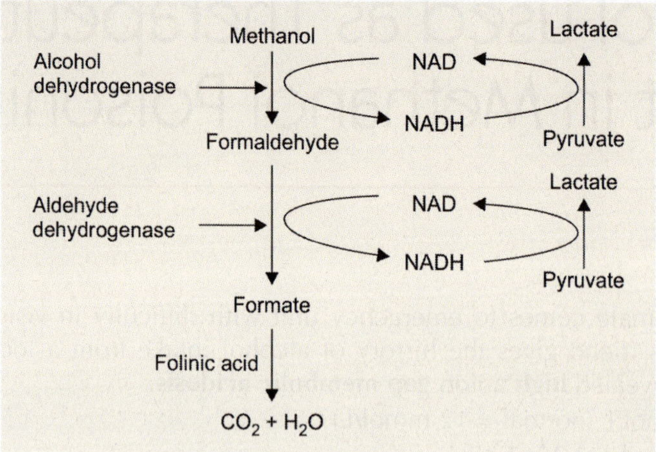

Fig. 1: Lactic acidosis in methanol poisoning

This **high level of NADH** thus produced tends to reduce pyruvate to lactate which results in lactic acidosis.

Bibliography

1. doi: 10.1186/1471-227X-10-3. PMID: 20175897; PMCID: PMC2836994.
2. Kumar P, Gogia A, Kakar A, Miglani P. An interesting case of characteristic methanol toxicity through inhalational exposure. J Family Med Prim Care. 2015 Jul-Sep;4(3):470–3. doi: 10.4103/2249-4863.161359. PMID: 26285665; PMCID: PMC4535118.

CASE

45

Scurvy in Adult

A 29-year-old male executive living alone presents with history of **bleeding gums** and **petechial hemorrhages*** on skin of lower limbs. Due to busy schedule, he is mainly dependent on frozen foods and frequent order of cheese pizza. His diet consisted of mainly carbohydrates which lacked fresh fruits and vegetables.

On examination, gums were swollen and slight bleeding near teeth were visible. Lower limb showed spontaneous **ecchymosis*. Perifollicular petechiae** were observed on trunk.

Blood vitamin C levels were found to be low (0.15 mg/dl)
(Normal levels of vitamin C: 0.20 to 2.00 mg/dl).

Vitamin C supplements were administered with dramatic improvement in the symptoms.

QUESTIONS

Q.1. What is the most probable diagnosis?

Q.2. What is the reason of bleeding gum in this person? Explain the role of vitamin C in the synthesis of collagen.

Q.3. Write two dietary sources rich in vitamin C.

Q.4. What is the treatment advised in this case?

Explanations

Ans.1. This patient is suffering with **scurvy**, a condition which arises due to deficiency of vitamin C (Fig. 1).

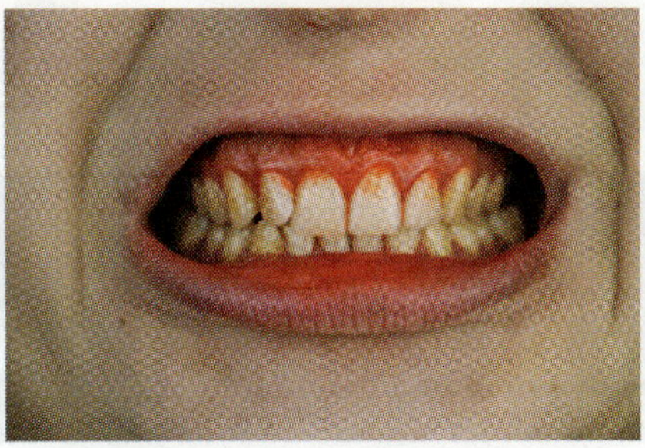

Fig. 1: Photo of scurvy gum

Ans.2. Vitamin C is necessary for synthesis of **collagen.** Collagen is important for blood vessel wall formation. Vitamin C is needed for action of **lysyl and prolyl hydroxylase enzyme** which is needed for conversion of proline and lysine to hydroxyproline and hydroxylysine. This conversion of amino acid is important to stabilize the structure of collagen.

In deficiency of vitamin C, structure of collagen is not stable and that results in weakening of blood vessel wall and fragility which is manifested in gums and also on skin as ecchymosis (Fig. 2).

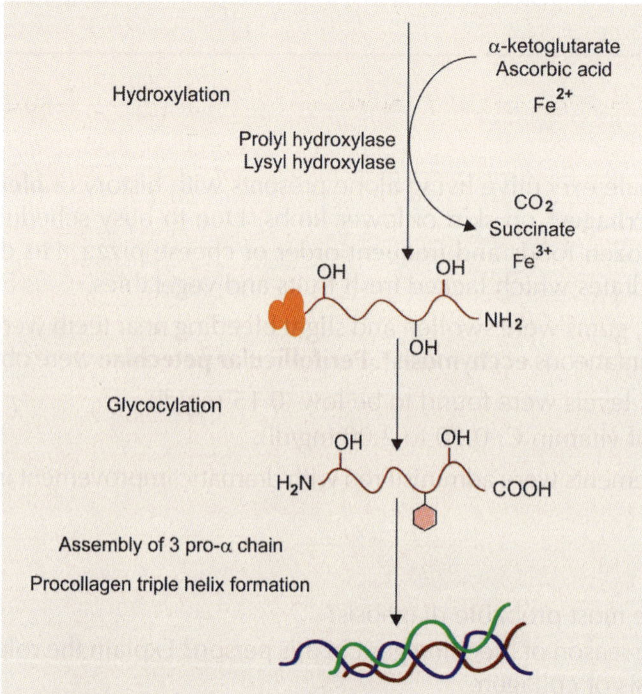

Fig. 2: Role of vitamin C in collagen synthesis

Ans.3. Most abundant source of vitamin C is **Indian Gooseberry (Amla) followed by guava and lime** (Fig. 3).

Fig. 3: Food items rich in vitamin C

Ans.4. This patient can be treated with supplementation of **1 g of vitamin C per day for five days followed by 500 mg per day for 2 weeks**.

In present era, scurvy is very rarely seen, though the individuals who are casual in their dietary habits and those who belong to lower socioeconomic strata, it is not very uncommon.

Bibliography

1. https://pmj.bmj.com/content/postgradmedj/43/502/539.full.pdf
2. Maxfield L, Crane JS. Vitamin C Deficiency. [Updated 2020 Jul 2]. In: StatPearls [Internet]. Treasure Island (FL): StatPearls Publishing; 2020 Jan-. Available from: https://www.ncbi.nlm.nih.gov/books/NBK493187/

Word Meaning

Petechial hemorrhages: Small pinpoint hemorrhages in skin or conjunctiva, which are result of minor bleed from capillary rupture.

Ecchymosis: It is discoloration of skin caused by leakage of blood in upper layer of skin. It is also known as bruise or purpura.

CASE

46

Scurvy in Child

A 3-year-old child brought by mother with complain of limpness, leg pain, lethargy, and irritability. This child is mainly taking milk and milk-containing food and is not fond of fruits and vegetables.

On examination, baby was **pallor** and there was **perifollicular hemorrhages** and **ecchymosis***. Lower limbs showing swelling over tibia regions. Oral examination revealed loose teeth and bleeding gums. X-ray lower limb showed **metaphyseal changes.**

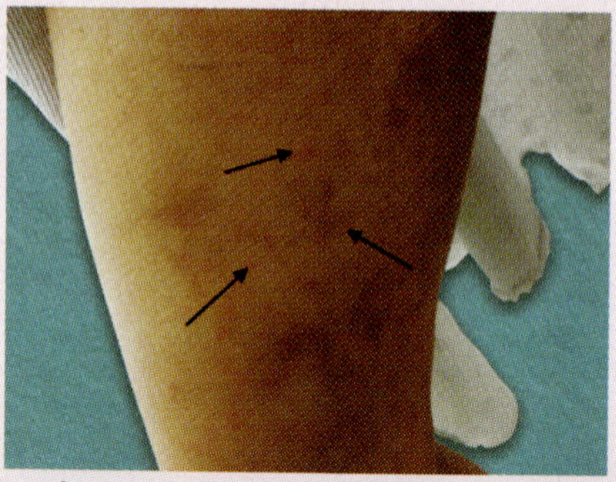

Petechial hemorrhages on lower limb due to fragile blood vessels

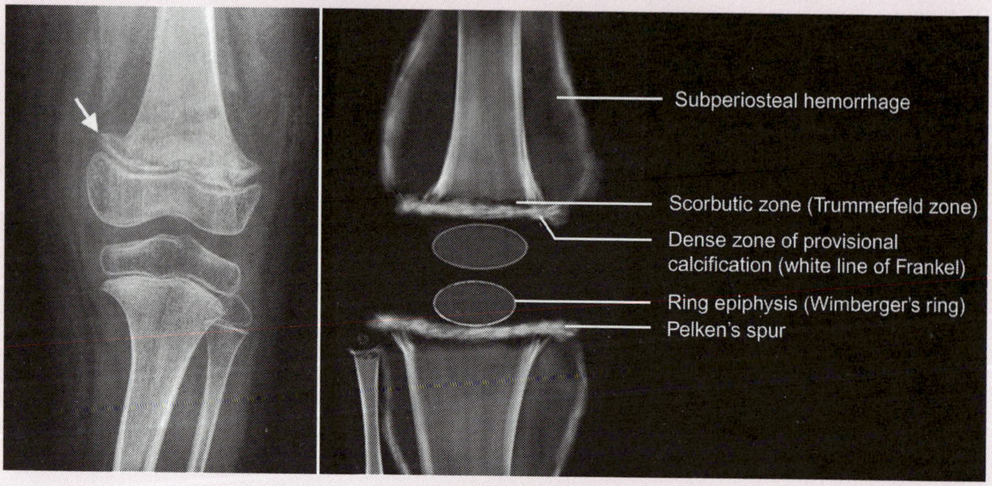

Metaphyseal changes

Following are the results of various laboratory investigations:
Low level of vitamin C as 0.12 mg/dl (normal levels: 0.20 to 2.00 mg/dl).
Peripheral smear showed **microcytic hypochromic anemia.**

QUESTIONS

Q.1. What may be the possibility of the above manifestation?
Q.2 What is the reason of bleeding gum, loose teeth and perifollicular hemorrhages?
Q.3. Why there is hypochromic microcytic anemia?
Q.4. What is infantile Barlow disease?
Q.5 What is the difference in bony changes in vitamin C and vitamin D deficiency?

Explanations

Ans.1. This baby is having **scurvy** as diagnosed by low level of vitamin C and clinical manifestation of petechiae and perifollicular hemorrhages.

Ans.2. As explained in the previous case, **formation of collagen is affected** due to lack of vitamin C. This results in weaking of blood vessel wall.

Ans.3. **Deficiency of vitamin C results reduced absorption of iron from gastrointestinal** tract resulting in **hypochromic microcytic anemia.** Vitamin C deficiency is a rare cause of anemia.

Ans.4. **Infantile Barlow disease** is the same as adult scurvy and is seen in infant aged 6 to 12 months. This disease is named after **Sir Thomas Barlow**. It is associated with bony **changes and mitral valve** involvement.

Ans.5. In vitamin C deficiency, **formation of osteoid matrix** is affected while in vitamin D deficiency osteoid matrix is synthesized normally and it is the mineralization of these osteoid matrices which are affected.

Bibliography

1. Leila GhediraBesbes, Samir Haddad, Chebil Ben Meriem, Mondher Golli, Mohamed-fadhel Najjar, Mohamed-NéjiGuediche, "Infantile Scurvy: Two Case Reports", *International Journal of Pediatrics*, vol. 2010, Article ID 717518, 4 pages, 2010. https://doi.org/10.1155/2010/717518

Word Meaning

Ecchymosis: Discoloration of skin caused by leakage of blood in upper layer of skin. It is also known as bruise or purpura.

CASE

47

Vitamin B_1 Deficiency (Wernicke-Korsakoff Psychosis)

A 56-year-old male shopkeeper presents with the history of forgetfulness, muscle weakness, poor appetite, and tremors in hand and nystagmus (involuntary eye movements). Attending physician noticed that he had unsteady gait when he entered the OPD and was looking confused.

History revealed that he is a known alcoholic for past 20 years taking 1 to 2 L of bear every day. On examination, he was found to have unsteady gait and fine tremors. He was found to have mental confusion too.

QUESTIONS

Q.1. What is the probable diagnosis in the above case?

Q.2. What nutrient deficiency may be suspected in the above case?

Q.3. What is the difference between wernicke's encephalopathy and Korsakoff's psychosis?

Q.4. How will you approach to this patient? What investigation would you require to confirm the diagnosis and what will be the treatment plan?

Explanations

Ans. 1 and 2. This seems to be a case of **"Wernicke-Korsakoff psychosis"** which is due to **moderately severe deficiency of vitamin B_1 (thiamine).** This is seen in chronic alcoholic and is due to poor diet intake and also due to impaired absorption of nutrients from gastrointestinal tract.

Mild deficiency of vitamin B_1 presents with **loss of appetite, fatigue, irritability, peripheral neuropathy** and is seen in elderly with poor diet intake.

Moderately severe deficiency is reflected in symptoms of **"Wernicke-Korsakoff psychosis"** described above. Wernicke disease is characterized by **clinical triad of GOA (global confusion, ophthalmoplegia, and ataxia)**. This classic triad is seen only in one-third of patients presenting clinically (Figs 1 and 2).

Ans. 3. This disease was described by two researchers independently in 19th century as **Wernicke's encephalopathy and Korsakoff's psychosis.** Finally, it was named Wernicke-Korsakoff syndrome.

Fig. 1: Nystagmus in alcoholic patient

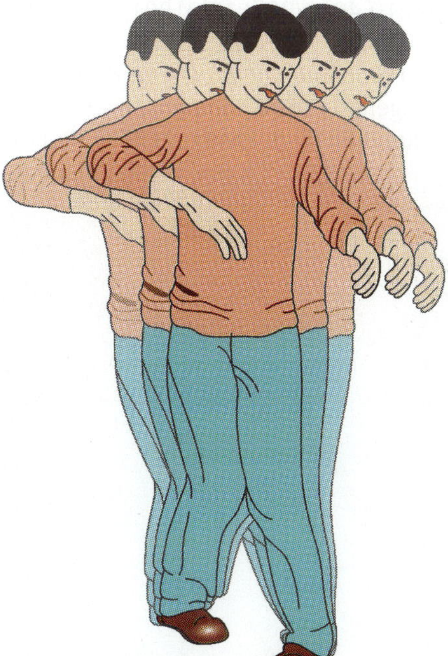

Fig. 2: Lack of coordination while walking

The two components of this disease, i.e. Wernicke's encephalopathy and Korsakoff's psychosis are seen in different phases of the same disease.

Wernicke's encephalopathy is acute neuropsychiatric manifestation of thiamine deficiency and should be treated immediately with parenteral thiamine supplementation to prevent progression to **Korsakoff's syndrome** which is characterized by **permanent brain damage and irreversible loss of memory.**

Korsakoff psychosis is the amnestic state where patient shows impairment of recent memory, though the past memory remains intact. **Confabulation** (falsification of facts) is an important characteristic where patient fills the memory gap by falsification which he believes to be true.

There is irreversible damage at **medial thalamic nucleic** and **mammillary bodies** in the brain. MRI shows **mammillary body atrophy** in chronic phase.

Ans. 4. **Vitamin B_1 level** can be estimated in plasma. **RBC transketolase activity** may be assessed to find out vitamin B1 deficiency. Patient may be given vitamin B_1 supplementation with other supportive therapy.

250 mg/day of thiamine intravenously along with **60 mg per oral twice a day** is prescribed till the clinical manifestation of this disease subsides.

Eye symptoms improve within weeks of starting the treatment and gait improves in months of such treatment.

Bibliography

1. Bilici R, Saridogan GE, Turan C, Goncu T, Akdur O, Citak S, Domac FM. A Case of Wernicke-Korsakoff Syndrome Treated 1 Year After the Onset of Symptoms. Prim Care Companion CNS Disord. 2015 Nov 26;17(6):10.4088/PCC.15l01801. doi: 10.4088/PCC.15l01801. PMID: 27057403; PMCID: PMC4805404.
2. Spiegel DR, Lim KJ. A case of probable Korsakoff's syndrome: a syndrome of frontal lobe and diencephalic structural pathogenesis and a comparison with medial temporal lobe dementias. Innov Clin Neurosci. 2011 Jun;8(6):15–9. PMID: 21779537; PMCID: PMC3140893.

CASE

48

Wet Beriberi (Shoshin Beriberi)

A 5-year-old male child of a poor labor presents with lethargy, poor appetite, irritability and pitting pedal edema. His family is mainly relying on diet consists of polished rice. Due to poverty, family is not able to afford fresh fruits and vegetables.

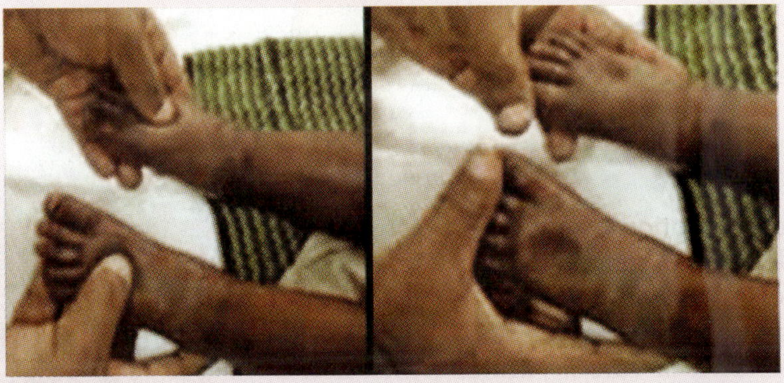

Pitting pedal edema

On examination, baby had pitting pedal edema in both the lower limbs. X-ray showed cardiomegaly* with pleural effusion*.

Blood lactic acid was found to be raised.

QUESTIONS

Q.1. What is the probable diagnosis in this case?

Q.2. What may the possible vitamin deficiency in this child?

Q.3. How the consumption of polished rice is related to the symptoms in this baby?

Q.4. Whether adults may present with wet beriberi or it is just the finding seen in children?

Explanations

Ans. 1 and 2. This baby is suffering with **wet beriberi (Shoshin beriberi)**. This disease is seen in severe case of vitamin B_1 (thiamine) deficiency. Thiamine is needed in **thiamine pyrophosphate** for many enzymatic reactions as coenzyme, mainly in carbohydrate metabolism where thiamine is importantly required for **PDH complex**

action, α-ketoglutarate dehydrogenase enzyme complex. Thus, they play very important role in production of ATP from glucose metabolism.

Beriberi may be:

a. **Dry beriberi**, characterized by neuromuscular symptoms, muscular atrophy and emaciation and peripheral neuropathy.
b. **Wet beriberi,** characterized by symptoms of cardiac involvement mainly right-sided heart failure presenting with breathlessness along with edema. Thiamine deficiency in long-term affects cardiac contractility landing up in heart failure.

Ans.3. Vitamin B_1 is found in **outer aleurone layer of cereals.** On polishing the rice, outer layer is removed and along with that even vitamin B_1 is lost (Fig. 1).

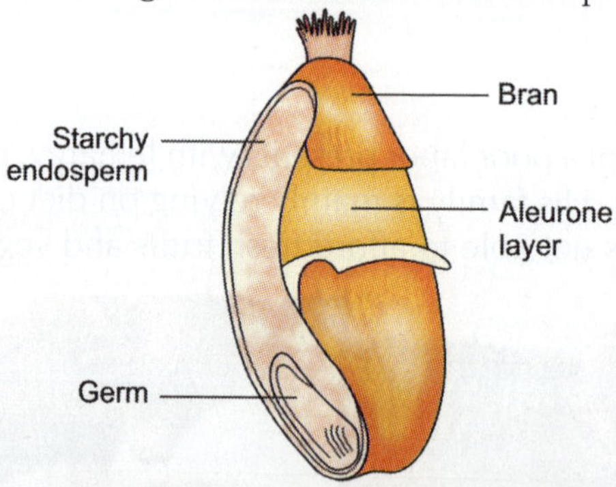

Fig. 1: Aleurone layer of cereal

Chronic consumption of polished rice results in vitamin B_1 deficiency, hence unpolished rice is more healthy to consume (Fig. 2).

This child has been consuming polished rice and mainly dependent of carbohydrates which is the reason of deficiency of vitamin B_1.

Fig. 2: Polished vs unpolished rice

Ans.4. It is not uncommon to find wet beriberi in adults who are deficient in thiamine either due to chronic malnutrition, alcoholism, etc. They may present with **heart failure, unexplained lactic acidosis,** etc., treatment with thiamine often improves the general condition.

Bibliography

1. Lee HS, Lee SA, Shin HS, Choi HM, Kim SJ, Kim HK, Park YB. A case of cardiac beriberi: a forgotten but memorable disease. Korean Circ J. 2013 Aug;43(8):569-72. doi: 10.4070/kcj.2013.43.8.569. Epub 2013 Aug 31. PMID: 24044018; PMCID: PMC3772304.
2. Lei Y, Zheng MH, Huang W, Zhang J, Lu Y. Wet beriberi with multiple organ failure remarkably reversed by thiamine administration: A case report and literature review. Medicine (Baltimore). 2018 Mar;97(9):e0010. doi: 10.1097/MD.0000000000010010. PMID: 29489643; PMCID: PMC5851725.

Word Meaning

Cardiomegaly: Enlargement of heart.

Pleural effusion: Accumulation of fluid in pleural space.

CASE

49

Ariboflavinosis (Vitamin B_2 Deficiency)

A 12-year-old boy presents with the history of repeated attack of **ulceration at the corner of mouth and swollen painful tongue.** He is son of a farmer and is pure vegetarian.

On examination, his **lips were dry and there were multiple ulcerations** at the corner of mouth which was making it difficult for him to open the mouth completely. Tongue was red and swollen.

He also gives the history of **photophobia** and difficulty in vision for past one year. Examination of eye showed **circumcorneal vascularization*.**

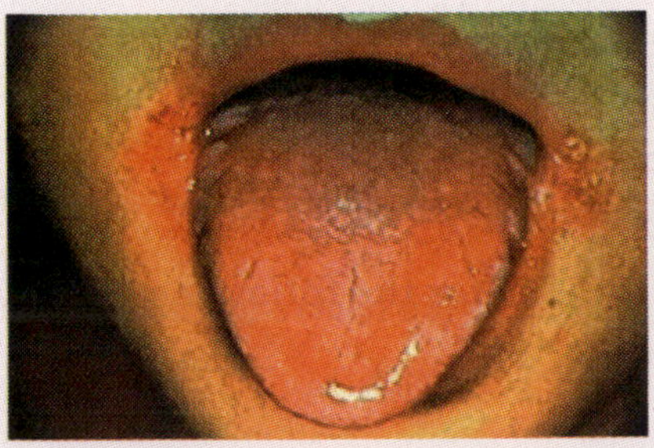

Swollen tongue and cracks at corner of mouth (ariboflavinosis)

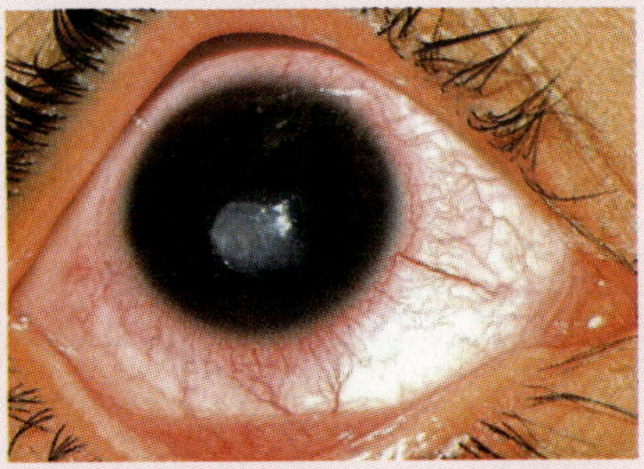

Circumcorneal vascularization

QUESTIONS

Q.1. What is the vitamin deficiency associated with these symptoms?

Q.2. What is the dietary source of this vitamin?
Q.3. What is the recommended daily allowance (RDA) of this vitamin?
Q.4. Mention coenzyme role of this vitamin.

Explanations

Ans.1. **Vitamin B_2 (riboflavin) deficiency is associated with above manifestation.** This condition is called **ariboflavinosis** and has been reported in older literatures in 40s.

B_2 deficiency is manifested in the form for **glossitis, angular stomatitis,** dry scaly lips, **circumcorneal vascularization,** and **bulbar conjunctiva proliferation.**

Such presentation of ariboflavinosis is not very common in today's era but is reported from time-to-time in country like India where poor strata of society heavily rely of carbohydrate-rich diet and have not access to milk and dairy products.

Ans.2. **Liver, yeast, egg, milk, fish, cereals are good sources of vitamin B_2.**

Ans.3. RDA of vitamin B_2 is 1.5 mg/dl.

Ans.4. Flavin adenine dinucleotide (FAD) and flavin mononucleotide (FMN) are coenzyme form of vitamin B_2.

FAD is required as coenzymes for the following enzymes:

a. Succinate dehydrogenase
b. Acyl-CoA dehydrogenase
c. Beta hydroxyacyl-CoA dehydrogenase
d. Cytochrome-b5 reductase

FMN is required as coenzyme for the following enzyme actions:

a. Cytochrome c reductase
b. L-amino acid oxidase

Bibliography

1. BK Das Gupta, Ariboflavinosis: With a Case Report On Parenchymatous Keratitis Following Riboflavin Deficiency, American Journal of Ophthalmology,1949;32(10)1389–93.
2. Deeny J. Riboflavin Deficiency: With a Case Report. Br Med J. 1942 Nov 21;2(4272):607. doi: 10.1136/bmj.2.4272.607. PMID: 20784550; PMCID: PMC2164705.

Word Meaning

Circumcorneal vascularization: Blood vessels around the cornea.

CASE

50

Pellagra (Niacin Deficiency)

A 35-year-old poor farmer on irregular diet presents with history of frequent diarrhea and **dry scaly lesions** around the neck for past 2 months. On examination, he was found to have characteristic lesions around the neck and dorsum of hand. Lesions were **itchy** specially on exposure to sun. He **grows maize and gives** the history of excess maize consumption for past 10 years.

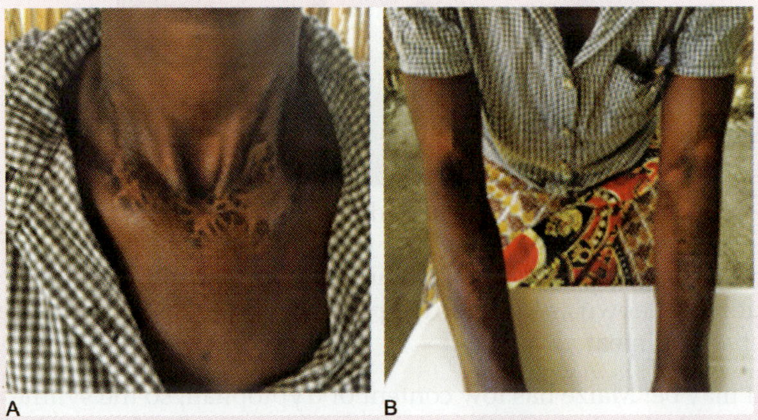

Characteristic Casal necklace dermatitis in neck region and scaly lesions on limb

QUESTIONS

Q.1. What is the diagnosis in above case and name the vitamin deficiency which may be associated with above symptoms?

Q.2. What is the **name given to this characteristic lesion** observed around the neck and on hand of the individual?

Q.3. Which amino acid may synthesize nicotinic acid *in vivo* in cell?

Q.4. Is the dependency of maize as staple diet may be a cause of this presentation? If yes, then why?

Q.5. Enlist the causes/conditions when patient may have pellagra or pellagra-like features.

Q.6. What is the treatment of this disease?

Explanations

Ans.1. This patient is suffering with pellagra (*pelle*: skin; *agra*: sour), a disease which is due to deficiency of vitamin B_3 (niacin or nicotinic acid).

This disease is characterized by 3 Ds: Diarrhea, dermatitis and dementia.

In extreme case which is left untreated, patient may die also. Pellagra was major cause of death until earlier 20th century and is now rarely seen in alcoholics, extremely poor population relying heavily on cheap source of carbohydrate like jowar, maize, etc.

Ans.2. Skin shows characteristic lesion around neck named **Casal necklace deformity and dry scaly lesions on hand known as gloved hand dermatitis.** The skin lesions in pellagra are quite characteristic where lesion starts in dorsum of hand as erythema associated with burning sensation and spreading upward involving forearm as well.

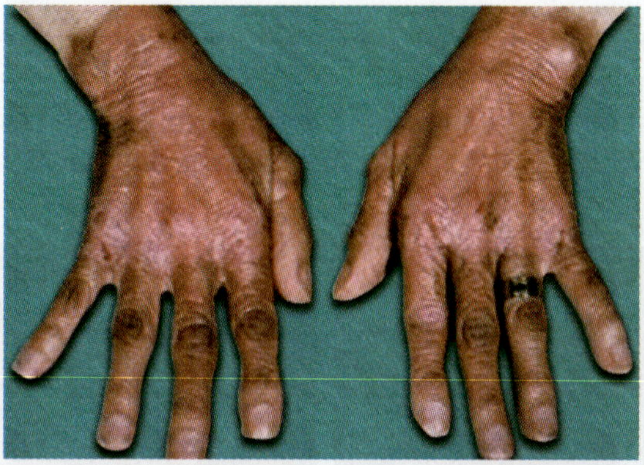

Fig. 1: Gloved hand dermatitis in pellagra

Ans.3. Tryptophan is amino acid which may produce **nicotinic acid in a vitamin B_6** requiring pathway known as **kynurenine pathway.** Tryptophan is abundantly found in legumes, fish and meat.

Ans.4. Yes it may be! Maize has low content of tryptophan, so the synthesis of niacin from tryptophan is deficient. On the other hand, the content of niacin is though high in maize, it is found in **bound form "Niacytin"**, hence not available for absorption.

Fig. 2: Excessive maize consumption may lead to pellagra

Similarly, another cereal which is associated with pellagra when consumed as staple diet is **jowar. Jowar has rich amount of L-leucine** which interferes with kynurenine pathway and hence reduces production of nicotinic acid from tryptophan. This is known as **"Leucine pellagra"**.

Ans.5. ***Causes of pellagra or pellagra-like features:***
- Niacin deficiency
- Excess corn/maize consumption
- Excess jowar consumption **(leucine pellagra)**
- Vitamin B_6 deficiency **(B_6 pellagra)**

Ans.6. Treatment of this disease is oral supplementation of vitamin B_3 in the dose **30 mg twice daily.**

Bibliography

1. Belavady B, Udayasekhara Rao P. Leucine and isoleucine content of jowar & its pellagragenicity. Indian J Exp Biol. 1979 Jul;17(7):659–61. PMID: 511231.
2. Carpenter KJ. The relationship of pellagra to corn and the low availability of niacin in cereals. Experientia Suppl. 1983;44:197–222. doi: 10.1007/978-3-0348-6540-1_12. PMID: 6357846.

CASE

51

B_6 Deficiency in Isoniazid (INH) Treatment

A 34-year-old engineer comes with complain of **tingling and numbness** in peripheries. His scalp is dry and scaly. He is a patient of pulmonary tuberculosis on treatment with antitubercular drugs for past 3 months.

QUESTIONS

Q.1. What vitamin deficiency will you suspect in this patient? Mention certain biochemical role of this vitamin.

Q.2. What is the RDA of concerned vitamin and what is the therapeutic dose of this vitamin in this case?

Q.3. What metabolite can be assessed in urine to find out the deficiency of vitamin B_6 in this individual and why?

Explanations

Ans.1. This patient is having symptoms of vitamin B_6 (pyridoxine) deficiency. **Isoniazid (INH)** is an antitubercular drug which is associated with vitamin B_6 deficiency.

Pyridoxine/vitamin B_6 forms a coenzyme pyridoxal phosphate (PLP). Reactions which require vitamin B_6 as PLP are mentioned below:

- Transamination
- Deamination
- Transsulfuration
- Condensation
- Tryptophan metabolism
- Decarboxylation

For carboxylation, biotin is required and for decarboxylation pyridoxine is required.

Vitamin B_6 is needed as coenzyme importantly in glycogen metabolism **(glycogen synthase and glycogen phosphorylase)** heme synthesis **(ALA synthase),** metabolism of sulfur-containing amino acid **(cystathionine β-synthase and cystathionase).**

Ans.2. RDA for vitamin B_6 is 1.5 mg/dl. Therapeutic dose is given in doses of 50–100 mg/dl per day.

Ans.3. Patients with vitamin B_6 deficiency excrete **xanthurenic acid** in the urine. Xanthurenic acid is the metabolite synthesized by tryptophan in alternate pathway of tryptophan metabolism when the normal route is impaired due to lack of activity of **kynureninase enzyme.**

This pathway is represented in Fig. 1.

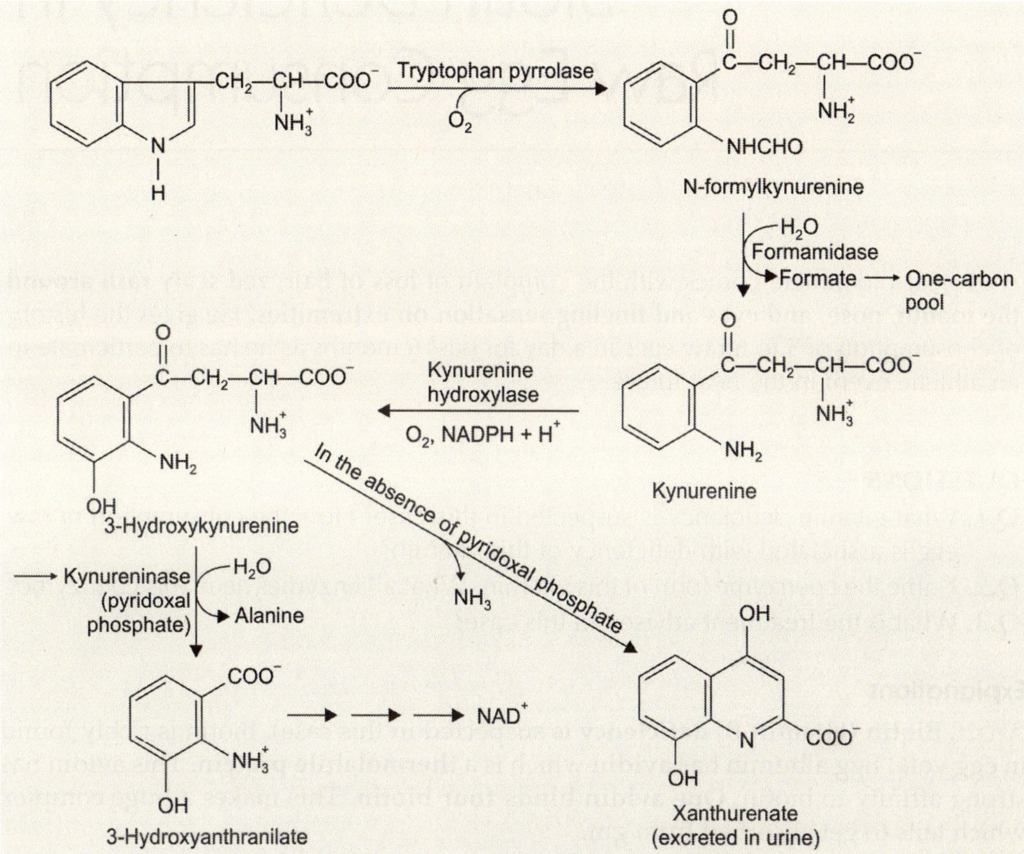

Fig. 1: Kynurenine pathway of tryptophan metabolism

As mentioned in Fig. 1, vitamin B_6 is required in form of PLP (pyridoxal phosphate) for the action of enzyme kynureninase. Lack of vitamin B_6 results in formation of xanthurenic acid in alternate pathway.

Bibliography

1. Bhargava M, Bhargava A. Pyridoxine for patients suffering from drug-susceptible tuberculosis in India. Public Health Action. 2018 Jun 21;8(2):97. doi: 10.5588/pha.18.0017. PMID: 29946528; PMCID: PMC6012958.
2. Snider DE Jr. Pyridoxine supplementation during isoniazid therapy. Tubercle. 1980 Dec; 61(4):191–6. doi: 10.1016/0041-3879(80)90038-0. PMID: 6269259.

CASE

52

Biotin Deficiency in Raw Egg Consumption

A 21-year-old athlete comes with the complain of **loss of hair, red scaly rash around the mouth, nose, and eyes and tingling sensation on extremities.** He gives the history of consumption of 5 to 6 raw eggs in a day for past 6 months as he has to participate in an athletic event in the near future.

QUESTIONS

Q.1. What vitamin deficiency is suspected in this case? How the consumption of raw egg is associated with deficiency of this vitamin?

Q.2. Name the coenzyme form of this vitamin. What all enzymes need this coenzyme?

Q.3. What is the treatment advised in this case?

Explanations

Ans.1. **Biotin (vitamin B_7 deficiency** is suspected in this case). Biotin is richly found in egg yolk. Egg albumin has **avidin** which is a **thermolabile protein.** This avidin has strong affinity to biotin. **One avidin binds four biotin.** This makes a huge complex which fails to get absorbed from gut.

Fig. 1: Egg white has avidin and egg yolk has biotin

Avidin-biotin binding characteristic is exploited in designing the enzyme conjugates in a technique called ELISA (enzyme-linked immunosorbent assay).

Consumption of raw egg for chronic duration results in deficiency of biotin. Boiling egg inactivates avidin and the biotin is absorbed from gut.

Sources of biotin are egg yolk, liver, cereals, spinach, mushrooms and rice.

Biotin deficient face, periorificial dermatitis, alopecia, and skin lesion resemble zinc deficiency lesions.

Ans.2. Coenzyme form of this vitamin is biocytin. Biotin acts as prosthetic group for carboxylase enzymes.

These carboxylases are:

- Pyruvate carboxylase (gluconeogenesis)
- β-methylcrotonyl-CoA carboxylase
- Propionyl-CoA carboxylase (amino acid metabolism)
- Acetyl-CoA carboxylase 1 and 2 (fatty acid synthesis)

Ans.3. Treatment with oral biotin supplementation is beneficial (5 mg/day).

Bibliography

1. Baugh CM, Malone JH, Butterworth CE Jr. Human biotin deficiency. A case history of biotin deficiency induced by raw egg consumption in a cirrhotic patient. Am J Clin Nutr. 1968 Feb; 21(2):173–82. doi: 10.1093/ajcn/21.2.173. PMID: 5642891.
2. https://www.sciencedirect.com/topics/biochemistry-genetics-and-molecular-biology/biotin-deficiency.

CASE

53

Vitamin B_{12} Deficiency in Vegans

A 24-year-old female who is strict vegetarian presents with history of easy tiredness, weakness and numbness sensation in extremities. On examination, she is pale and peripheral smear shows **megaloblastic cell** with **hypersegmented neutrophils.**

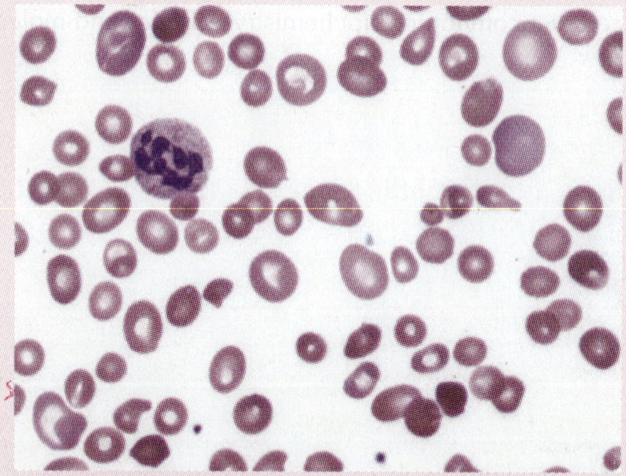

Megaloblastic anemia with hypersegmented neutrophils

Following are the results of various laboratory investigations:
Vitamin B_{12}: 20 pmol/L (normal = 120–700 pmol/L)
Hb: 7.5 g% (normal = 12–15 g%)
MCV: 102 fl (normal = 83–101 fl)

QUESTIONS

Q.1. Which vitamin deficiency may be suspected in the above case?
Q.2. What enzymes need vitamin B_{12} as cofactor?
Q.3. How much of this vitamin is stored in human?
Q.4. What is the mechanism of absorption of vitamin B_{12}?
Q.5. What is the treatment in this case?

Explanations

Ans.1. **Deficiency of vitamin B_{12} is** suspected in this case as the patient is purely vegetarian and vitamin B_{12} is obtained only from foods of animal origin like meat, egg, fish and dairy products.

Finding of **hypersegmented neutrophils** with five or more segmented lobe is characteristic finding seen in B_{12} or folic acid deficiency.

Ans.2. Vitamin B_{12} is **cobalt containing vitamin** needed for two important enzymes:
- **Methylmalonyl-CoA mutase (B_{12} needed as 2-deoxyadenosyl (ado) cobalamin**
- **Methionine synthase (B_{12} needed in methylcobalamin form)**

Ans.3. Vitamin B_{12} is stored in human. **Total 2 to 3 mg of vitamin B_{12} is stored in human body**. This store lasts for 2 to 3 years since the intake of vitamin B_{12} totally stopped.

RDA is 1 to 3 μg per day.

Ans.4. Vitamin B_{12} (cobalamin) is **absorbed from ileum. Gastric intrinsic factor** which is secreted by **gastric parietal cell** in fundus and body of stomach is important for absorption of vitamin B_{12}.

One intrinsic factor binds two molecules of cobalamin. This IF-cobalamin complex enters the ileal cell via receptor known as **"cubilin".** Within the ileal cell IF is destroyed and cobalamin is absorbed in portal circulation where it binds with transcobalamin II (TC II).

In pernicious anemia, autoantibodies develop against intrinsic factor which results in severe deficiency of intrinsic factor, hence absorbtion of vitamin B_{12} is adversely affected.

Ans.5. Injection of **1000 μg hydroxycobalamin** may be given once a week for 5 weeks.

Bibliography

1. Annelli L, Watterson R, Pandya R. *et al.* Vitamin B_{12} deficiency with combined hematological and neuropsychiatric derangements: a case report. *J Med Case Reports* 8, 277 (2014). https://doi.org/10.1186/1752-1947-8-277.
2. Rizzo G, Laganà AS, Rapisarda AM, La Ferrera GM, Buscema M, Rossetti P, Nigro A, Muscia V, Valenti G, Sapia F, Sarpietro G, Zigarelli M, Vitale SG. Vitamin B_{12} among Vegetarians: Status, Assessment and Supplementation. Nutrients. 2016 Nov 29;8(12):767. doi: 10.3390/nu8120767. PMID: 27916823; PMCID: PMC5188422.

CASE

54 Vitamin B_{12} Deficiency Presenting as Neurological Manifestation

A 62-year-old male presents with history of difficulty in walking, frequent fall, postural dizziness, and easy tiredness for past 6 months. Family members complain that he has become highly irritable and does not take care of himself and has mood swing for past 6 months. He had partial gastric resection 20 years back.

Following are the results of various laboratory investigations:
Vitamin B_{12}: 18 pmol/L (normal = 120–700 pmol/L)
Hb: 8.4 g% (normal = 12–15 g%)
MCV: 98 fl (normal = 83–101 fl)
Peripheral smear shows megaloblastic picture with hypersegmented nuclei.

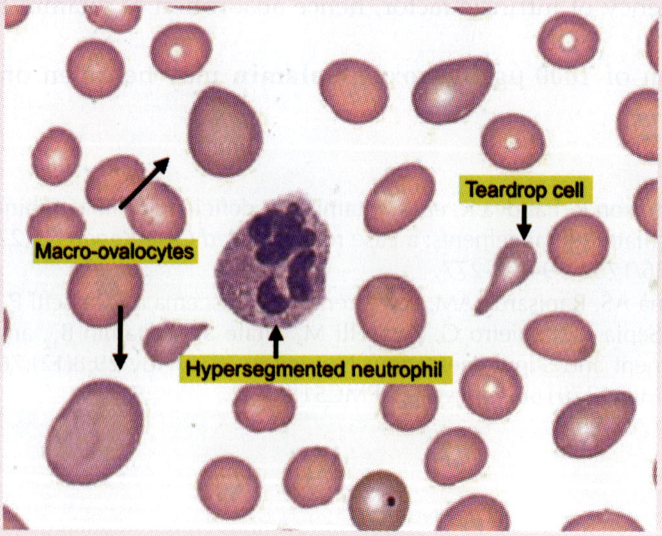

Megaloblastic anemia

QUESTIONS
Q.1. What may be the diagnosis in this case?
Q.2. Is the history of gastric resection is important in this patient?
Q.3. What is the treatment of this patient?

Explanations

Ans.1. This patient has **neurological involvement along with anemia in vitamin B_{12} deficiency.** Vitamin B_{12} is required for **nerve sheath myelination.** Deficiency of B_{12} results in demyelination of posterior and lateral tracts of spinal cord in cervical and thoracic region. Cranial nerves may also be involved.

Ans.2. Yes it is important. **Gastric parietal cells secrete intrinsic factor** which is important for absorption of vitamin B_{12}. In patients with gastric resection, the absorption of vitamin B_{12} does not take place and they should be given IM injections of vitamin B_{12} lifelong.

Ans.3. Treatment is difficult as the neurological manifestations are mostly irreversible. **Hydroxycobalamin injection is given 1000 µg per day twice a week for 6 doses.**

Bibliography

1. Serin HM, Arslan EA. Neurological symptoms of vitamin B_{12} deficiency: analysis of pediatric patients. Acta Clin Croat. 2019 Jun;58(2):295–302. doi: 10.20471/acc.2019.58.02.13. PMID: 31819326; PMCID: PMC6884369.
2. Voukelatou P, Vrettos I, Kalliakmanis A. Neurologic symptoms as the only manifestation of B_{12} deficiency in a young patient with normal hematocrit, MCV, peripheral blood smear and homocysteine levels. Oxf Med Case Reports. 2016 Dec 1;2016(12):omw091. doi: 10.1093/omcr/omw091. PMID: 28031855; PMCID: PMC5184828.

CASE

55 Vitamin B_{12} Deficiency Presenting as Knuckle Hyperpigmentation

A 29-year-old female comes to dermatologist with case of **hyperpigmentation of knuckles** which she has been noticing for past 2 years. She is pure vegetarian and gives the history of easy tiredness on doing daily chores.

Her blood examination showed following picture: Hypersegmented neutrophil with macro-ovalocyte.

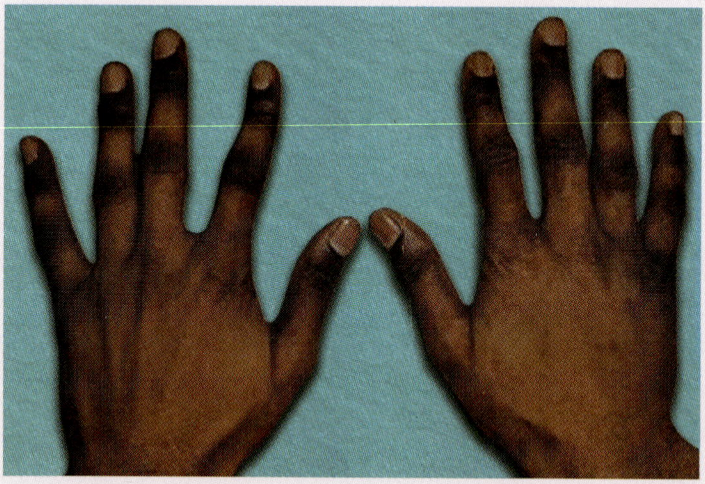

Following are the results of various laboratory investigations:

Hemoglobin: 7.6 g% (normal = 12–15 g%)

Mean corpuscular volume: 111 fl (normal = 83–101 fl)

WBC (TLC): 3.5×10^9/L (normal = $4–10 \times 10^9$/L)

Platelet count: 240 thousand/µl (normal = 150–410 thousand/µl)

Peripheral blood film showed oval macrocytes, hypersegmented neutrophils and teardrop RBC.

The serum vitamin B_{12} level was 95 pmol/L (normal = 120–700 pmol/L).

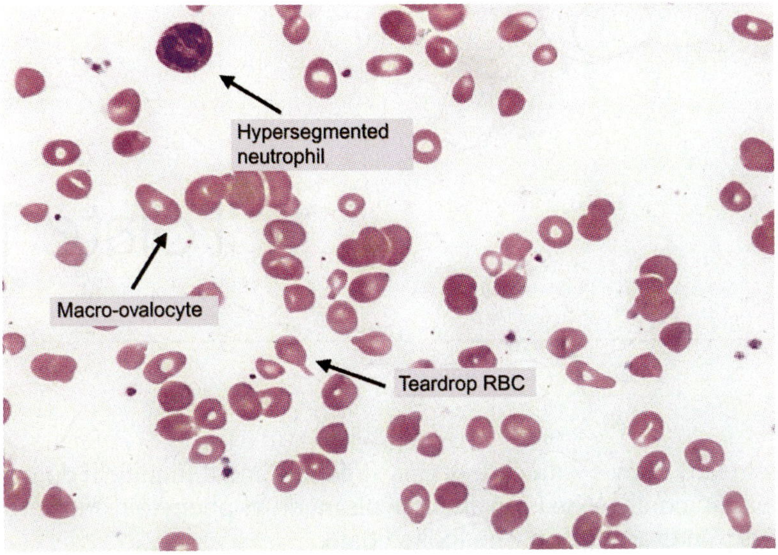

QUESTIONS

Q.1. What may be possible cause for this presentation?
Q.2. In what other conditions, hyperpigmentation of knuckle is seen?
Q.3. How will you confirm the diagnosis?
Q.4. What is the treatment advised for this case?

Explanations

Ans.1. Around 10% of vitamin B_{12} deficiency patients show reversible knuckle hyperpigmentation and pigmentation of oral mucosa.

Ans.2. Other causes of knuckle hyperpigmentation

- PCOD (polycystic ovarian disease)
- Acanthosis nigricans
- Raynaud's disease
- Addison's disease
- Dermatomyositis

Ans.3. Assessment of vitamin B_{12} deficiency can be done by measuring its plasma level. Normal level of vitamin B_{12} in the plasma is 120–700 pmol/L. Any value lesser than this indicates vitamin B_{12} deficiency. Vitamin B_{12} is also known as animal protein as it is found in food of animal origin.

Ans.4. Supplementation of vitamin B_{12} is advised.

Bibliography

1. Agrawala RK, Sahoo SK, Choudhury AK, Mohanty BK, Baliarsinha AK. Pigmentation in vitamin B_{12} deficiency masquerading Addison's pigmentation: A rare presentation. Indian J Endocr Metab 2013;17, Suppl S1:254–6.
2. Mori K, Ando I, Kukita A. Generalized hyperpigmentation of the skin due to vitamin B_{12} deficiency. J Dermatol. 2001 May;28(5):282-5. doi: 10.1111/j.1346-8138.2001.tb00134.x. PMID: 11436369.

CASE

56

Folate Trap

A 35-year-old lady comes with case of easy tiredness and difficulty in doing household chores for past 4 months. She is on anticonvulsant drugs phenytoin and barbiturates for past 10 years. On examination, she looked pale.

Blood investigation **revealed B_{12} and folic acid deficiency.** Urine showed the presence of **FIGLU (formiminoglutamic acid). Peripheral smear** shows presence of **megaloblastic cells.**

QUESTIONS

Q.1. What may the condition patient is suffering from?

Q.2. What is folate trap?

Q.3. What is FIGLU and in which condition FIGLU gets excreted in urine?

Explanations

Ans.1. The clinical presentation and blood, urine findings denote that this patient is suffering with vitamin B_{12} deficiency which causes **folate trap.**

Ans.2. **Deficiency of B_{12} does not allow the action of methionine synthase enzyme for which cofactor N5-methyltetrahydrofolic acid is required. This reaction is**

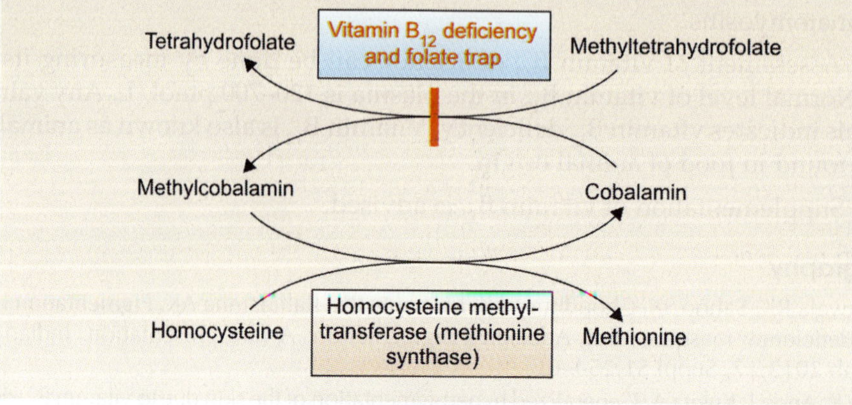

Fig. 1: Vitamin B_{12} deficiency and folate trap

important to supply free form of THF which is then converted to various forms of THF like N^5, N^{10} methylene, THF, etc. used in diverse sets of biochemical reactions.

Deficiency of vitamin B_{12} does not allow this reaction to go on and results in deficiency of free THF. This condition is known as "folate trap".

Ans.3. Formiminoglutamic acid (FIGLU) is the intermediate of histidine metabolism. Its further metabolism requires folic acid. In deficiency of folic acid, FIGLU gets excreted in the urine as it is not further metabolized by the body. The metabolism of histidine and excretion of FIGLU is described in Fig. 2

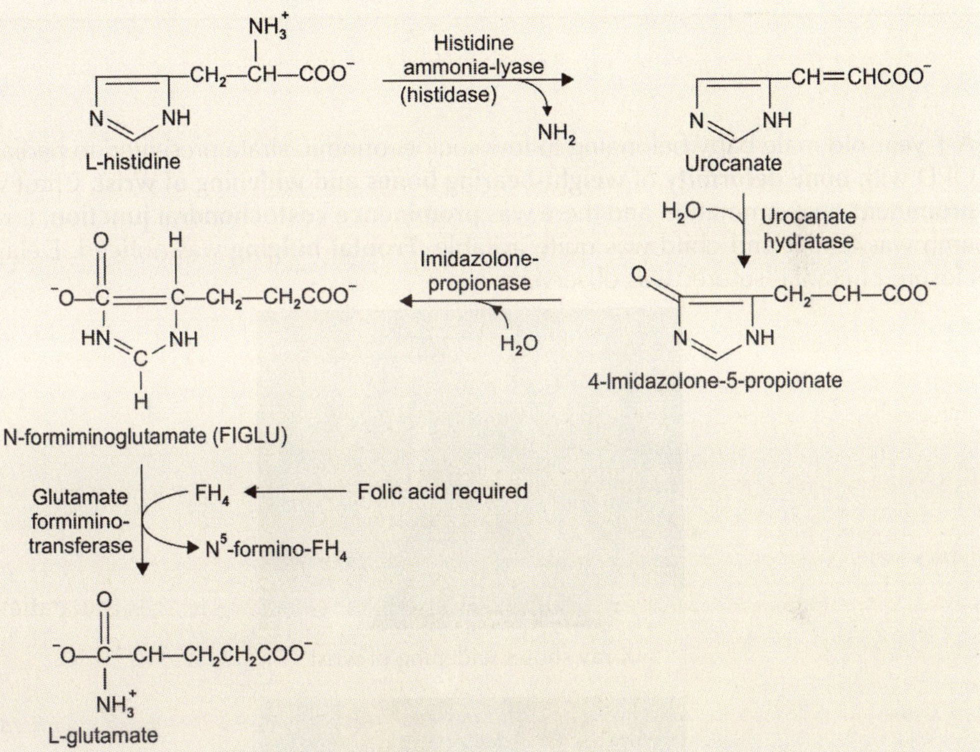

Fig. 2: Metabolism of histidine which requires folic acid (FH_4)

Bibliography

1. Hoffbrand AV, Jackson BF. Correction of the DNA synthesis defect in vitamin B_{12} deficiency by tetrahydrofolate: evidence in favour of the methyl-folate trap hypothesis as the cause of megaloblastic anaemia in vitamin B_{12} deficiency. Br J Haematol. 1993 Apr;83(4):643–7. doi: 10.1111/j.1365-2141.1993.tb04704.x. PMID: 8518179.
2. Scott JM, Weir DG. The methyl folate trap. A physiological response in man to prevent methyl group deficiency in kwashiorkor (methionine deficiency) and an explanation for folic-acid induced exacerbation of subacute combined degeneration in pernicious anaemia. Lancet. 1981 Aug 15;2(8242):337–40. doi: 10.1016/s0140-6736(81)90650-4. PMID: 6115113.

CASE

57

Rickets

A 1-year-old male baby belonging to low socioeconomic strata presented to pediatric OPD with **bone deformity** of **weight-bearing bones and widening of wrist.** Chest was **prominent** on examination and there was **prominence costochondral junction.** Lower limb was tender and child was quite irritable. **Frontal bulging** was noticed. Delayed closure of cranial sutures was observed.

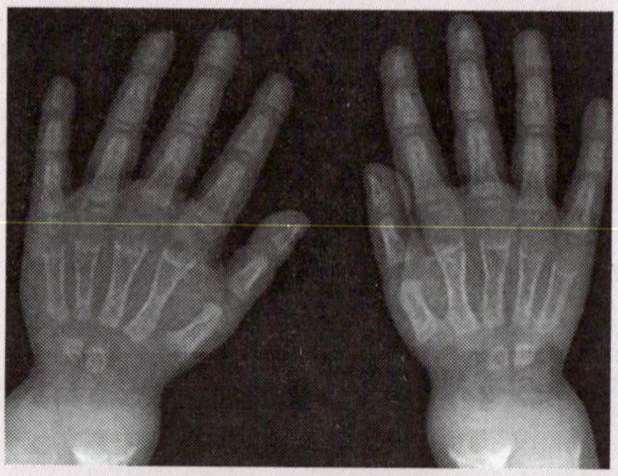

X-ray shows widening of wrist

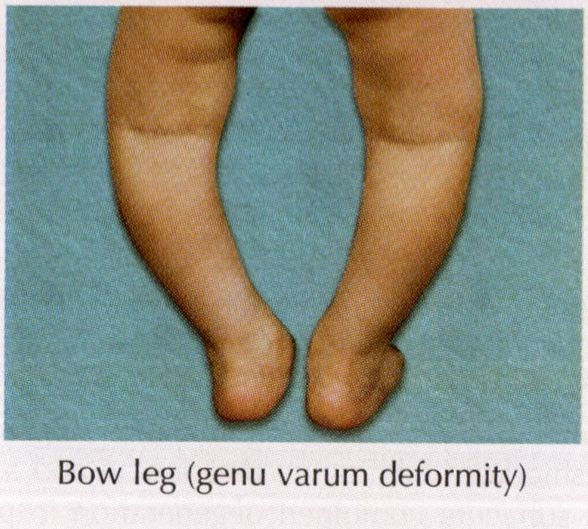

Bow leg (genu varum deformity)

Following are the results of various laboratory investigations:

Serum calcium: 7.5 mg/dl (normal = 8.5–10.5 mg/dl)

Serum phosphate: 2.7 mg/dl (normal = 3.5–4.5 mg/dl)

ALP: 675 IU/L (normal = 98–279 IU/L)
PTH was also found to be elevated.
Level of 25-hydroxycholecalciferol was normal = 20 ng/ml (normal = 30–50 ng/ml)
Level of 1,25-dihydroxycholecalciferol was undetectable.

QUESTIONS

Q.1. What may be the possible diagnosis in the above case?

Q.2. Why the level of most potent vitamin D 1,25-dihydroxycholecalciferol was undetectable?

Q.3. Which micronutrient deficiency is responsible for such manifestations?

Q.4. What type of lower limb deformity seen in rickets? **Genu valgum (knocked knee) or genu varum (bowed leg)?**

Q.5. What is the treatment plan for this baby?

Explanations

Ans.1. This child is suffering with **deficiency of vitamin D,** which results in a disease known as **rickets**. Rickets is a condition which is seen in children who have **deficiency of vitamin D before the fusion of epiphyseal plate** which results in **growth retardation** and **widening of growth plate.**

Deficiency of vitamin D may be due to **reduced intake, reduced synthesis of vitamin D in skin, increased urinary loss, reduced activation** or may be due to **resistance of 1α-hydroxycholecalciferol.**

This condition is very common in India specially in children belonging to poor socioeconomic strata. Young child of 4 to 6 years mainly suffers with this disease. In children before the fusion of epiphyseal plate, deficiency of vitamin D results in **expansion (widening) of growth plate in long bones (bowed legs) as well as in costochondral junction (rachitic rosary).**

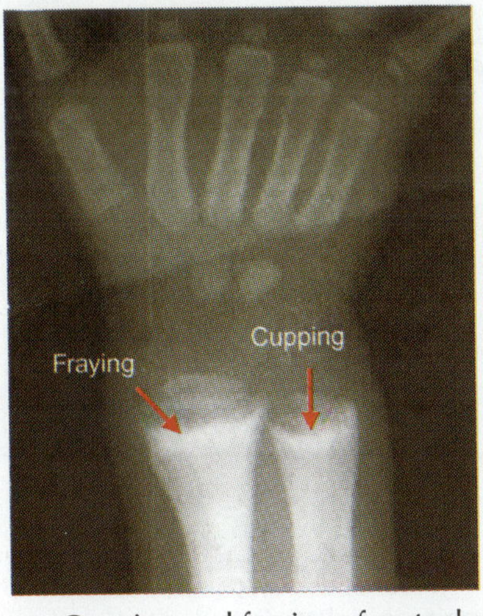

Fig. 1: Cupping and fraying of metaphysis

Long-standing vitamin D deficiency results in hypocalcemia along with features of secondary hyperparathyroidism. Bone mineralization gets affected manifesting in decreased bone mineral density and osteopenia in X-ray.

Rickets may be broadly classified into two subtypes:

a. **Pseudovitamin D deficiency rickets (PDDR)/vitamin D-dependent rickets type 1 (VDDR-1)**
b. **Vitamin D resistance rickets**

Mutation of gene for 1α-hydroxylase results in a disorder termed **pseudovitamin D deficiency rickets (PDDR)/vitamin D-dependent rickets type 1 (VDDR-1).** It is an **autosomal recessive** disorder which presents early within 1 year of life with growth retardation, rickets and hypocalcemic seizures. Level of 1,25 $(OH)_2D$ is low while the level of 25-hydroxycholecalciferol is normal. PTH is high. Therapy with vitamin D metabolite is helpful.

Another is **vitamin D resistance rickets** where there occurs mutation of vitamin D receptor. It is difficult to treat. Level of PTH is high and also the level of 1,25 $(OH)_2D$ is high. Clinical presentation is same as in vitamin D-dependent (PDDR) but **alopecia** may be **an additional finding in vitamin D resistance rickets.**

It is difficult to treat and daily infusion of calcium and phosphate may be desired.

Both above types of rickets can be differentiated biochemically by level of 1,25 $(OH)_2D$ which is undetectable in former and is higher in latter.

Radiological Finding in Rickets

Widening and expansion of growth plate of long bones, costochondral junction (rachitic rosary).

Ans.2. This child is suffering with **pseudovitamin D deficiency rickets (PDDR)/** vitamin D-dependent rickets type 1 (VDDR-1) which will have undetected level of 1,25$(OH)_2D$ as explained avove.

Ans.3. The main reason behind these manifestations of vitamin D deficiency is hypocalcemia due to impaired calcium absorption from intestine. This results in impaired mineralization of bone, manifesting as **osteopenia** as explained above.

Ans.4. May be either genu varum or genu valgum or may be a mixed presentation known as **'windswept deformity'** (Fig. 2).

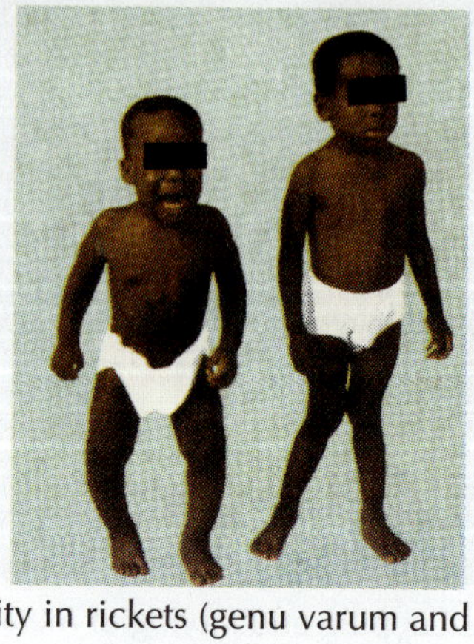

Fig. 2: Lower limb deformity in rickets (genu varum and genu valgum respectively)

Ans.5. Vitamin D is given in therapeutic doses, along with physiotherapy will be advised.

In vitamin D deficiency, proximal myopathy is common in both children and adults.

Bibliography

1. Füchtbauer L, Brusgaard K, Ledaal P, Frost M and Frederiksen AL (2015). Case report: vitamin D-dependent rickets type 1 caused by a novel CYP27B1 mutation. *Clinical case reports*, *3*(12), 1012–1016. https://doi.org/10.1002/ccr3.406
2. KebashniThandrayen, John M. Pettifor, The roles of vitamin D and dietary calcium in nutritional rickets, Bone Reports, Volume 8, 2018, Pages 81–89, ISSN 2352-1872.
3. Lemoine E, Giabicani V, Lockhart E, Grimprel P, Tounian, Case report of nutritional rickets in an infant following a vegan diet, Archives de Pédiatrie, Volume 27, Issue 4, 2020, Pages 219–222.

CASE

58

Osteomalacia Case 1

A 55-year-old postmenopausal female comes with complain of backache and lower limb pain. On examination, generalized bone tenderness was noticed and kyphosis was seen. X-ray long bones showed **porous matrix and Looser's zone.** She reported two episodes of fracture small bones of foot in last 3 years. She has been on restricted food free of dairy products and minimal sun exposure for past 4 years.

Following are the results of various laboratory investigations:

ALP: 1453 U/L (normal = 98–278 U/L)

PTH: 654 pg/ml (normal = 10–65 pg/ml)

Total serum calcium: 7.9 mg/dl (normal = 8.5–10.5 mg/dl)

Phosphorus: 3.1 mg/dl (normal = 3.5–4.5 mg/dl)

25-hydroxycholecalciferol: 7 ng/ml (normal = 30–50 ng/ml)

QUESTIONS

Q.1. What may be the probable diagnosis in this case?

Q.2. What may be the cause of this disease?

Q.3. Which age group is affected by above disease and why?

Q.4. What is Looser's zone on X-ray?

Explanations

Ans.1. This woman is **having osteomalacia** as obvious in X-ray finding **where loss of mineralization of osteoid matrix is seen.** Hypocalcemia and hypophosphatemia which accompany vitamin D deficiency results in impaired **mineralization** of bone matrix protein. This type of bone which is deficient in mineralization is weaker than normal bone and is prone for bowing in weight-bearing extremities and also is prone for fractures. Severe vitamin deficiency also causes proximal myopathy in both children and adults.

Typical biochemical hallmarks of osteomalacia include **low serum calcium, low phosphate, high levels of parathyroid hormone (PTH), and high levels of alkaline phosphatase and osteocalcin along with low levels of 25-hydroxycholecalciferol.**

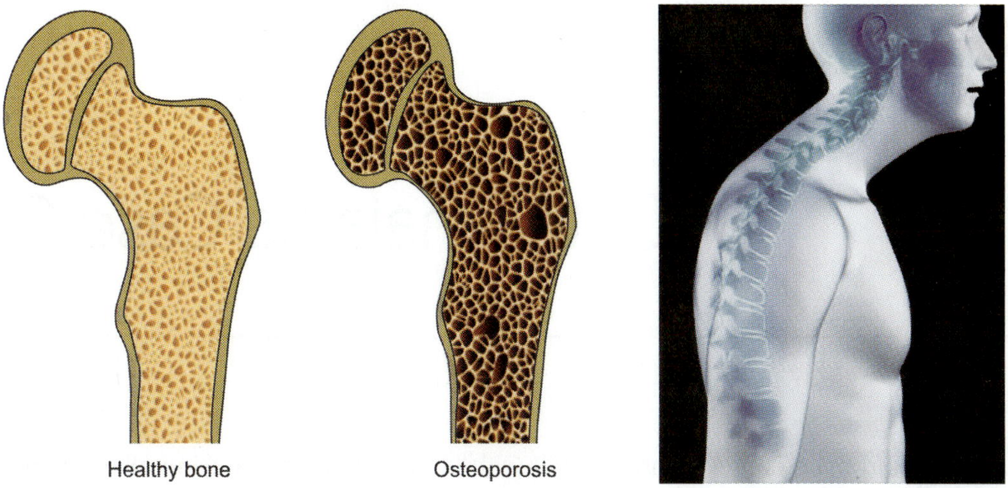

Fig. 1: Porous bone in osteoporosis and kyphosis

Ans.2. Elderly population specially postmenopausal women are prone to this disorder.

Ans.3. **Pseudofractures or Looser's zones** are specific radiological finding in osteomalacia seen in scapula, pelvis and femoral neck. These are **thin radiolucent lines** which are many centimeter long and are formed because of arterial pulsations over the bone.

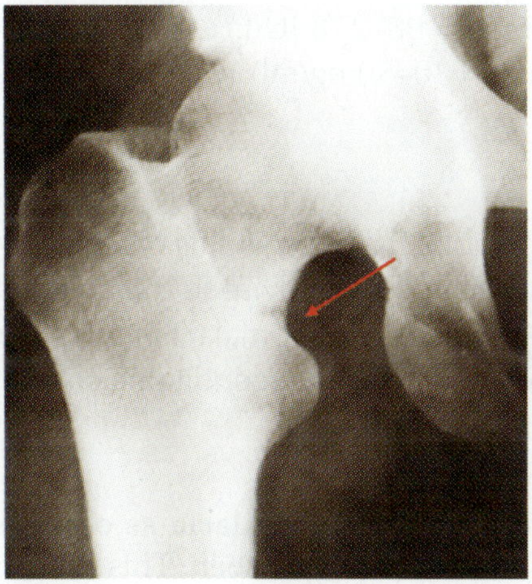

Fig. 2: Looser's zone

Bibliography

1. Marco Infante, Diana Corradini, Maria Romano, Alfonso Bellia, and Davide Lauro (*2018*) Life-threatening Hypocalcemia in a Patient With Highly Suspected Osteomalacia: A Case Report. AACE Clinical Case Reports: March/April 2018, Vol. 4, No. 2, pp. e174-e178.
2. Shikino K, Ikusaka M, Yamashita T. Vitamin D-deficient osteomalacia due to excessive self-restrictions for atopic dermatitis. BMJ Case Rep. 2014 Jul 4;2014:bcr2014204558. doi: 10.1136/bcr-2014-204558. PMID: 25100811; PMCID: PMC4091129.

CASE

59

Osteomalacia Case 2

A 60-year-old male comes to ortho OPD with complain of **diffuse generalized bone pain** and difficulty in getting up from sitting position for past 6 months. On examination, **kyphosis (forward bending)** was observed. X-ray long bones showed deficient mineralization of osteoid matrix and presence of **Looser's zone** in femoral neck. He reported two episodes of fracture small bones of foot in last 3 years.

Biochemical investigation revealed low serum calcium, phosphorus, and vitamin D (25-hydroxycholecalciferol) and raised ALP levels.

Serum calcium: 8.5 mg/dl (normal = 8.5–10.5 mg/dl)

Serum phosphorus: 3 mg/dl (normal = 3.5–4.5 mg/dl)

Serum ALP: 240 IU/L (normal = 98–278 IU/L)

Vitamin D: 9 ng/ml (normal = 30–50 ng/ml)

QUESTIONS

Q.1. What may be the probable diagnosis in this case?

Q.2. What other investigation will you advise in this case?

Q.3. What is the difference between osteomalacia and osteoporosis?

Q.4. How will you differentiate these two conditions clinically and biochemically?

Explanations

Ans.1. This man is suffering with **osteomalacia** as obvious in X-ray finding where **loss of mineralization of osteoid matrix is seen.** This condition is due to deficiency of vitamin D.

Hypocalcemia and hypophosphatemia which accompany vitamin D deficiency results in impaired mineralization of bone matrix protein. Hypomineralized matrix results in bowing of weight-bearing extremities and **skeletal fractures.** Severe vitamin deficiency also causes **proximal myopathy** in both children and adults. Osteomalacia in children is specifically named **rickets.**

Pseudofractures or Looser's zones (Milkman line) are specific finding in osteomalacia seen in scapula, pelvis and femoral neck. They are seen as **transverse lucencies** with sclerotic borders traversing pathway through a bone, usually perpendicular to the involved cortex.

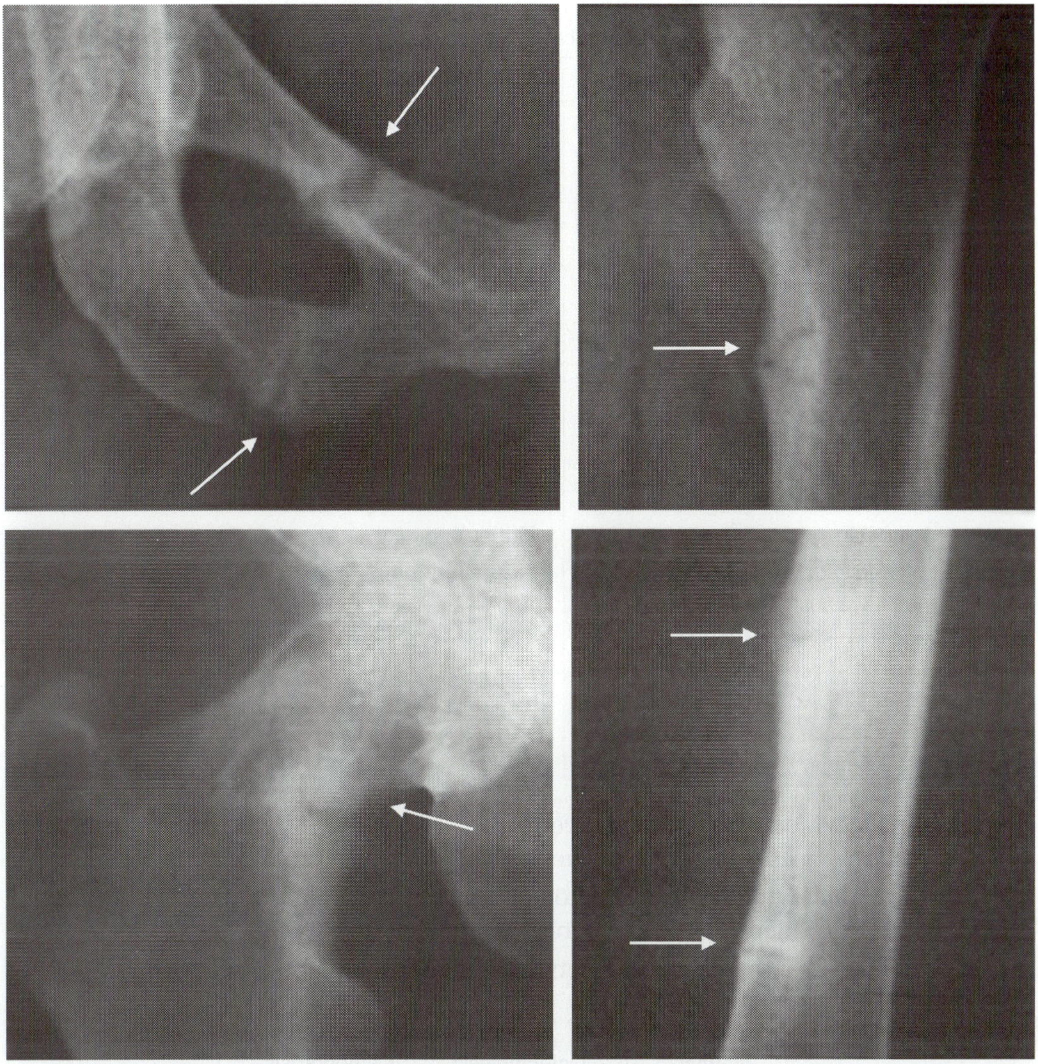

Fig. 1: Looser's zone as transverse lucencies with sclerotic border

Ans.2. BMD **(bone mineral density)** can be advised which scans calcium and phosphorus in bone.

Ans.3. In **osteoporosis,** bone is **porous** and **fragile** while in **osteomalacia,** it is **pliable and soft** and tends to bend.

Ans.4. Osteoporosis tends to remain asymptomatic until patient develops a fracture. On the other hand, in osteomalacia, clinical presentation is bone pain, generalized musculoskeletal weakness and repeated fracture.

Biochemically osteoporosis is characterized by relatively normal calcium, phosphorus level while in case of osteomalacia level of calcium and phosphorus is low.

Salient features of difference in osteomalacia and osteoporosis are described in Table 1.

Table 1: Clinical/biochemical and radiological difference between osteomalacia and osteoporosis

Osteomalacia	*Osteoporosis*
Due to deficiency of vitamin D either due to deficient sun exposure, dietary deficiency, malabsorption, liver of kidney failure. All age (in child it is called rickets)	In postmenopausal women due to lack of Estrogen Steroid intake Old age, postmenopausal female Patient on chronic steroid therapy Oophorectomy
Defect in bone matrix (osteoid) mineralization	Bone absorption is more and new synthesis is less
Bone matrix synthesized normally	Less matrix synthesis
Pliable soft bone which bends	Fragile bone which breaks
Generalized bone pain and muscle weakness	Generally asymptomatic till fracture occurs. Pain only after fracture
Treatment possible	Difficult to treat
Looser's zone on X-ray	No such finding
Biochemically Low calcium, low phosphorus	Biochemically
High ALP	Normal level of calcium, phosphorus and
High PTH	ALP
Ca × P = <2.4	Ca × P = >2.4

In osteomalacia, osteoid matrices are formed normally, it is the mineralization of these osteoid matrices which are deficient due to lack of vitamin D.

On the other hand, in osteoporosis, too little bone is formed but whatever is formed is normally mineralized. This condition is due to lack of osteoblastic activity.

Figure 2 illustrates the difference between osteomalacia and osteoporosis.

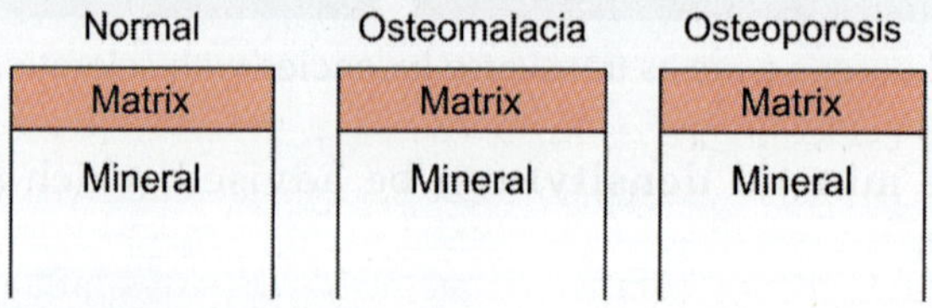

Fig. 2: Difference in osteomalacia and osteoporosis (in osteomalacia mineralization is deficient. In osteoporosis, available bone is normal)

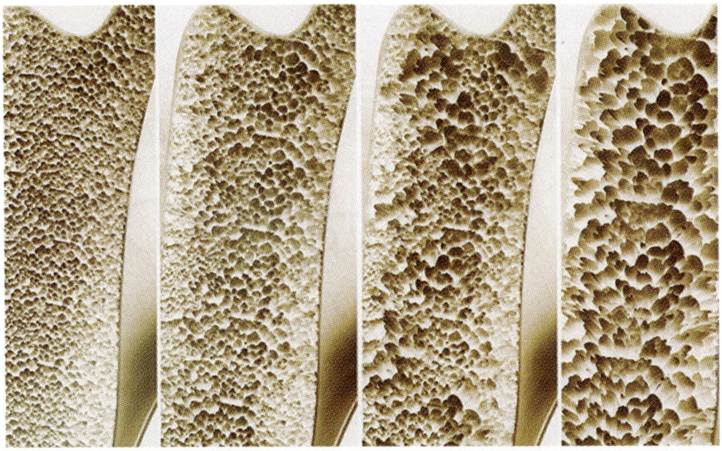

Fig. 3: Osteoporosis of long bones, increasing level of severity

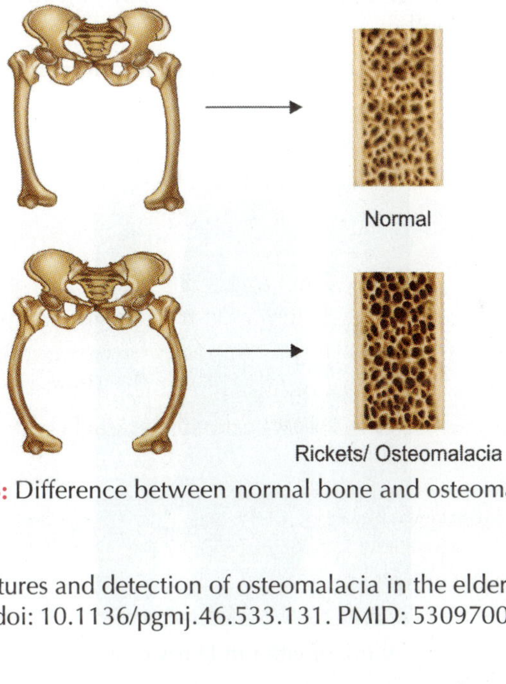

Fig. 4: Difference between normal bone and osteomalacia

Bibliography

1. Rosin AJ. Clinical features and detection of osteomalacia in the elderly. Postgrad Med J. 1970 Mar;46(533):131–6. doi: 10.1136/pgmj.46.533.131. PMID: 5309700; PMCID: PMC2466969.

CASE

60

Vitamin D Toxicity

A 6-year-old male child comes with the history of severe headache, vomiting, **bulging fontanelle.** History revealed that the child was treated by a quack for growth retardation, who prescribed him **vitamin D capsule 60,000 IU,** which the child has been taking twice a week for past 6 months.

X-ray showed **metastatic calcification of calf muscle and renal tract stones.**

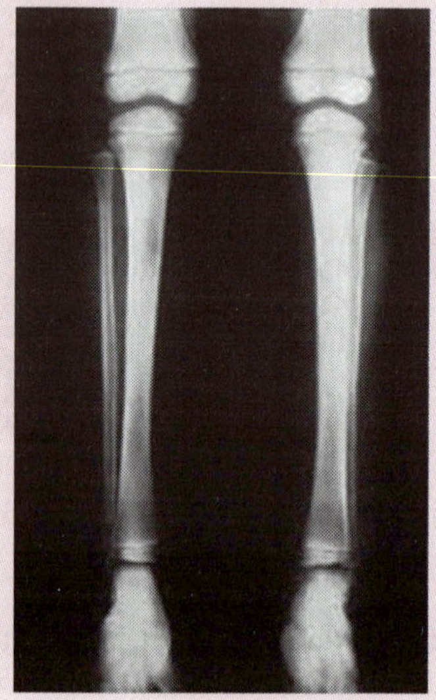

X-ray of vitamin D toxicity

Following are the results of various laboratory investigations:

Calcium: 13.6 mg/dl (normal = 8.5–10.5 mg/dl)

25-hydroxycholecalciferol (calcidiol): 45 ng/ml (normal = 30–50 ng/ml)

QUESTIONS

Q.1. What may be the cause of above presentation?

Q.2. What is the toxic dose of vitamin D? What will be the status of bone mineralization in this toxicity?

Explanations

Ans.1. This child has **vitamin D toxicity** due to excessive consumption of vitamin D.

Bone will be demineralized just like seen in rickets, but this time it is due to excess reabsorption of calcium from the bone rather than deficiency of mineralization of the bone. **High plasma level of 1,25(OH)$_2$D and calcium level are important finding in vitamin D toxicity.**

Classical presentation is anorexia, irritability, abdominal pain, polydipsia, polyuria, hypercalciuria, nephrocalcinosis.

Diagnosis of vitamin D toxicity is confirmed by high serum calcium, high serum vitamin D 25(OH) and normal PTH.

Treatment consists of stopping of vitamin D intake and avoiding all calcium-containing food in diet (such as milk, curd, etc.).

Ans.2. Upper tolerable limit for vitamin D toxicity is 4000 IU/day in adults and in children it is 2000 IU/day. In addition to vitamin D, another vitamin which is toxic in excess dose is vitamin A.

Bibliography

1. Maji D. Vitamin D toxicity. Indian J Endocrinol Metab. 2012 Mar;16(2):295-6. doi: 10.4103/2230-8210.93773. PMID: 22470872; PMCID: PMC3313753.
2. Marins TA, GalvãoTde F, Korkes F, Malerbi DA, Ganc AJ, Korn D, Wagner J, Guerra JC, Borges Filho WM, Ferracini FT, Korkes H. Vitamin D intoxication: case report. Einstein (Sao Paulo). 2014 Apr;12(2):242-4. doi: 10.1590/s1679-45082014rc2860. PMID: 25003934; PMCID: PMC4891171.

CASE

61

Vitamin A Deficiency

A 10-year-old female child presents with history of **decreased vision after sunset** for past 1 year. Mother gives the history of repeated upper respiratory infection in the child. On examination, she was found to have **follicular hyperkeratosis** on the back of the arm and white patches on outer sclera in both the eyes.

QUESTIONS

Q.1. What micronutrient deficiency may be the reason of such presentation in this child?

Q.2. What may be the reason of follicular hyperkeratosis in this child?

Q.3. What is phrynoderma?

Q.4. What is the name given to the eye manifestation?

Q.5. What dietary advice to be given for this child?

Explanations

Ans.1. This child is suffering with vitamin A deficiency.

Ans.2. People deficient in this vitamin are prone for repeated infection and follicular hyperkeratosis.

This vitamin is responsible for downregulation of keratin synthesis. In deficiency of vitamin A, there is excessive keratinization of skin and mucous membrane. In skin, it shows up as **follicular hyperkeratosis** and in mucous membrane of respiratory, gastrointestinal and genitourinary tract such keratinization favor colonization of microorganism and infection (Fig. 1).

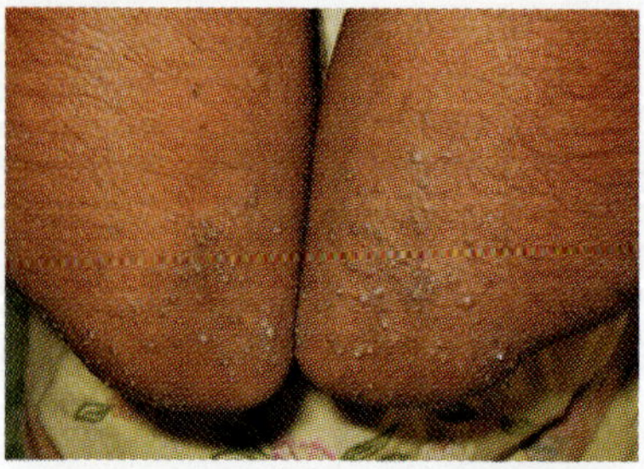

Fig. 1: Follicular hyperkeratosis

Ans.3. Phrynoderma is the distinct form of follicular hyperkeratosis. Phrynoderma is believed to be associated with severe malnutrition. Various literatures show deficiency of vitamins E, B, A and essential fatty acid being associated with phrynoderma, but general malnutrition seems to be the strongest correlation with phrynoderma.

Ans.4. In the eye, the specific name given to these white patches are **Bitot's spots** which are patches of keratinized epithelium on the sclera (Fig. 2).

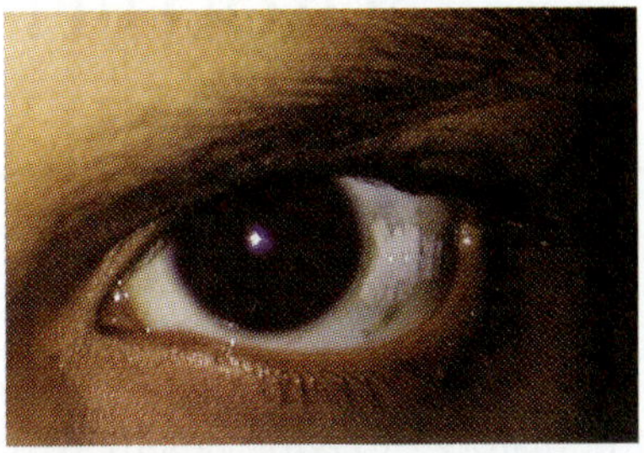

Fig. 2: Bitot's spot

Ans.5. Yellow and green vegetables have rich amount of vitamin A, e.g. mango, banana, carrot, tomato, green leafy vegetable, pumpkin. Vitamin A can be given in therapeutic doses to this child (Fig. 3).

Fig. 3: Food items rich in vitamin A

Therapeutic dose of vitamin A 7,500 to 35,000 units intramuscularly once a day. Duration of therapy: 10 days.

Bibliography

1. Cheshire J, Kolli S. Vitamin A deficiency due to chronic malabsorption: an ophthalmic manifestation of a systemic condition. Case Reports 2017;2017:bcr-2017-220024.
2. Chiu M, Watson S. Xerophthalmia and vitamin A deficiency in an autistic child with a restricted diet. BMJ Case Rep. 2015 Oct 5;2015:bcr2015209413. doi: 10.1136/bcr-2015-209413. PMID: 26438671; PMCID: PMC4600805.

CASE

62

Vitamin A Toxicity

An 8-year-old child presents with case of headache, anorexia, pruritus, and a lack of weight gain. Mother gives the history of giving health syrup for 6 months to the child advised by a quack in the village as child had **difficulty in vision in the night.** X-ray finding—notice hyperostosis in the middle of the shaft of long bones.

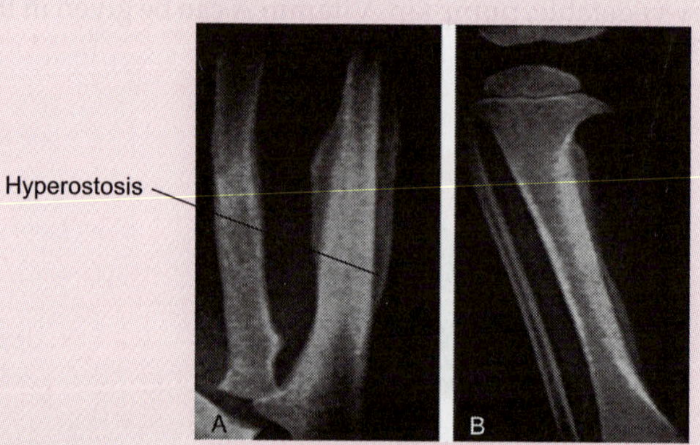

QUESTIONS

Q.1. What may be the possible cause of such presentation in the child?

Q.2. What is the clinical presentation in **acute and chronic toxicity of vitamin A**?

Q.3. What dose of vitamin A is reported to be responsible for acute and chronic toxicity, respectively?

Q.4. What is the advice given to mother?

Explanations

Ans.1. This child is showing the symptoms and signs of **chronic toxicity of vitamin A** which is the consequence of excessive ingestion of vitamin A in past six months.

Ans.2. Acute as well as chronic toxicity of vitamin A has been reported from time to time. Reason for toxicity in excess dose is obscure and possible reason may be oversaturation of its binding protein.

Acute hypervitaminosis: Vitamin A toxicity has occurred in infants in developing countries after ingestion of very high amounts of vitamin A during vaccine administration. It has also been reported in adults after consumption of diet rich in vitamin A, e.g. "**ocean perch liver" consumption**.

Common symptoms include nausea, vomiting, and drowsiness, numbness, pruritus*, facial flushing*, desquamation of skin and less common symptoms include diplopia*, papilledema*, cranial nerve palsies, and other symptoms suggestive of pseudotumor cerebri. **Nephrocalcinosis** and liver toxicity have also been reported.

Chronic hypervitaminosis: It results from excessive ingestion of vitamin A for several weeks or months. Signs of subacute or chronic toxicity may include **headache; vomiting; anorexia; dry, itchy desquamating skin; seborrheic cutaneous lesions; fissuring at the corners of the mouth; alopecia and/or coarsening of the hair; bone abnormalities.**

Radiographs show hyperostosis affecting several long bones, especially in the middle of the shafts.

Ans.3. The acute toxic dose of vitamin A is 25,000 IU/kg and the chronic toxic dose is 4,000 IU/kg every day for 6–15 months.

Ans.4. No specific antidote is available to treat vitamin A toxicity. Supplementation, if any, taken by the patient should be immediately stopped. Symptomatic management is advised with proper **hydration** to the patient.

Bibliography

1. Chiu YK, Lai MS, Ho JC, Chen JB. Acute fish liver intoxication: report of three cases. Changgeng Yi Xue Za Zhi. 1999 Sep;22(3):468-73. PMID: 10584420.
2. Homma Y, Otani N, Ishimatsu S. A Case Report of Acute Vitamin A Intoxication due to Ocean Perch Liver Ingestion. J Emerg Med. 2015 Jul;49(1):15-7. doi: 10.1016/j.jemermed.2014.12.056. Epub 2015 Apr 4. PMID: 25850632.
3. Refractory hypercalcemia owing to vitamin A toxicity in a 4-year-old boy. Melissa Lorenzo, Maxime Nadeau, Jennifer Harrington, Peter J. Gill. CMAJ Jun 2020, 192 (25) E671-E675; **DOI:** 10.1503/cmaj.191568

Word Meaning

Pruritus: Itching

Facial flushing: Redness of face

Diplopia: Double vision

Papilledema: Edema of optic disc

CASE

63 Vitamin K Deficiency in Newborn

In a labor room, 27-year-old female delivered a male baby. Attending pediatrician immediately gave vitamin K injection IM after taking care of skin and respiration. Intern on duty was asked to explain the rationale behind giving the vitamin K injection.

QUESTIONS

Q.1. What is the indication of vitamin K injection to all newborn babies?
Q.2. What is the reason behind nomenclature of vitamin K as vitamin K?
Q.3. How many types of vitamin K are known? Which form of vitamin K is injected?
Q.4. Which form of vitamin K is used in ETC?
Q.5. How to assess deficiency of vitamin K in a person?
Q.6. Name the food item rich in vitamin K.
Q.7. What is the RDA of vitamin K as per current ICMR guidelines?
Q.8. Explain vitamin K cycle (epoxide cycle).

Explanations

Ans.1. **0.5 to 1 mg of vitamin K injection** should be given to all newborn within 24-hour of birth. Infants born to mothers taking drugs like **anticonvulsants** (carbamazepine, phenytoin and barbiturates), **antituberculosis drugs** (isoniazid, rifampicin), some **antibiotics** (cephalosporins) and **vitamin K antagonists** (coumarin, warfarin) are prone to develop vitamin K deficiency bleeding, if they are not supplemented with vitamin D soon after birth.

The clinical presentation is often severe with cephalic hematoma and intracranial and intra-abdominal hemorrhages are not very rare. This is known as **hemorrhagic disease of newborn (HDN)/vitamin K deficiency bleeding (VKDB).**

Ans.2. Vitamin K was discovered in 1930 by biochemist **Henrik Dam.**

Alphabet **"K" stands for coagulation**. Vitamin K is important for blood coagulation and hence it is known as coagulant vitamin. Vitamin K is required for **post-translational modification** of clotting factor II (prothrombin), factors VII, IX, and X, protein C, protein S, and protein Z. Other proteins which are post-translationally modified are bone metabolism proteins like osteocalcin, periostin and **matrix Gla protein** as well as vascular biology protein like growth-arrest-specific gene 6 protein.

CASE

64

Iron deficiency Anemia

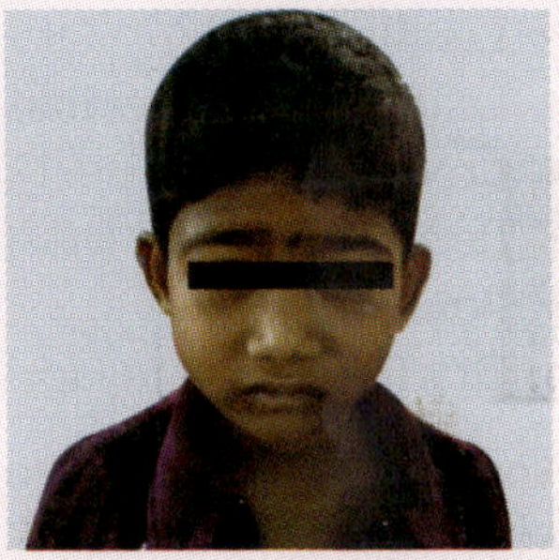

A 6-year-old boy comes to the OPD with complain of loss of appetite, early fatigue and lethargy*. Mother complained that boy gets tired very soon while playing and suffer with headache frequently. On examination, nails were flat and brittle and pallor of face was seen.

Following are the results of various laboratory investigations:

Hb: 9.0 g/dl

Serum iron: 41 ng/dl (normal = 50–150 ng/dl)

Serum ferritin: 40 µg/dl (normal = 50–200 µg/dl)

TIBC (total iron binding capacity): 470 µg/dl (normal = 250–450 µg/dl)

MCV: 75 mm^3/fl (normal = 80–95 mm^3/fl)

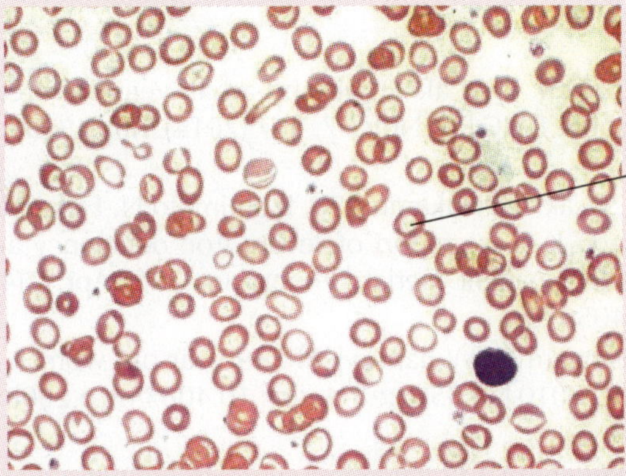

Peripheral smear shows microcytic hypochromic picture

QUESTIONS

Q.1. What may be the probable diagnosis?

Q.2. What are the important causes which will lead to this disorder in Indian scenario?

Q.3. What dietary compounds have rich amount of iron?

Q.4. What is the role of iron in hemoglobin synthesis?

Explanations

Ans.1. This child is suffering with **iron deficiency anemia** which is characterized by **microcytic hypochromic picture of peripheral smear.** Lack of iron impairs heme synthesis resulting in microcytic hypochromic picture. Early tiredness is due to inability of RBC to supply sufficient oxygen to tissues.

Ans.2. Main cause of iron deficiency in Indian scenario is **dietary deficiency** of iron, superimposed by **high phytate content in Indian diet** which impair absorption of iron by making insoluble complex with iron. In addition, **worm infestation** is a major cause of iron deficiency in Indian population.

Ans.3. Diets having rich amount of iron are:

- Jaggery
- Beans
- Lentils
- Leafy green vegetable like spinach
- Broccoli
- Red meat pork
- Poultry
- Dry fruits: Raisin, apricot

Ans.4. Hemoglobin is composed of heme and globin. Synthesis of heme needs iron which is incorporated in the center of protoporphyrin ring with the help of enzyme ferrochelatase.

In deficiency of iron this terminal step of heme synthesis is affected and there occurs deficiency of heme which affects production of hemoglobin adversely.

Bibliography

1. Bock F, Borucki K, Vorwerk P, et al. A two-and-a-half-year-old breastfed toddler presenting with anemia: a case report. BMC Res Notes 7, 917 (2014). https://doi.org/10.1186/1756-0500-7-917
2. Mabuchi S, Suzuki R, Sasaki M, Nakamura M, Izumimoto N, Hakamada T, Toyoshima K, Abe Y, Kaneko E, Shimokado K. Case report of severe iron deficiency anemia caused by proton pump inhibitor in an elderly patient. GeriatrGerontol Int. 2017 Apr;17(4):662–663. doi: 10.1111/ggi.12908. PMID: 28405975.
3. Palmer SR, Thanarajasingam G, Wolanskyj AP. 39-year-old woman with an obscure case of anemia. Mayo Clin Proc. 2010 Jan;85(1):e1–4. doi: 10.4065/mcp.2008.0721. PMID: 20042553; PMCID: PMC2800282.

Word Meaning

Lethargy: Loss of energy and enthusiasm.

CASE

65

Hemosiderosis, Bronze Diabetes, Bantu Siderosis

A 60-year-old diabetic, nonsmoker and nonalcoholic male comes with complain of brown discoloration of skin of back and arm. He has no history of blood transfusion and is not on iron pills. He deny the cooking of food in iron vessels. This patient was diagnosed to have primary hemochromatosis on liver biopsy*.

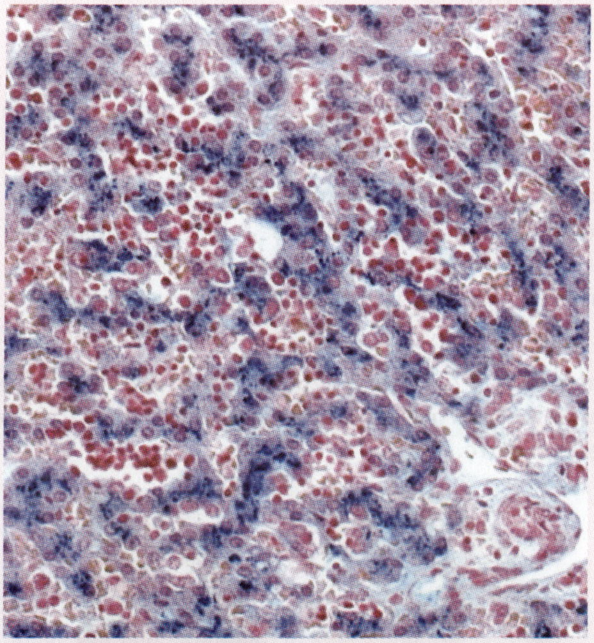

Liver shows hemosiderosis

QUESTIONS

Q.1. What are the causes of iron overload?

Q.2. What is the difference between hemochromatosis and hemosiderosis?

Q.3. What is bronze diabetes?

Q.4. What is Bantu siderosis?

Explanations

Ans.1. There are two causes of iron overload:

a. Primary iron overload (hereditary hemochromatosis)
b. Secondary iron overload

Primary Iron Overload

Primary iron overload is inherited condition which is autosomal recessive (AR) and is due to increased iron absorption (up to ten times than normal rate) from the intestine.

This results in **deposition of iron in parenchymal cells,** particularly in the liver, pancreas, heart; endocrine glands like thyroid and pituitary, and synovium.

Secondary Iron Overload

Secondary iron overload is because of following reasons:

i. ***Iron-loading anemias:*** Thalassemia, sideroblastic anemia, sickle cell disease, chronic hemolytic anemia, aplastic anemia, pyruvate kinase deficiency.
ii. ***Chronic liver diseases:*** Hepatitis C infection, nonalcoholic fatty liver disease, alcoholic fatty liver disease.
iii. ***Iatrogenic causes:*** Red blood cell transfusion, long-term hemodialysis.

In secondary iron overload, **iron accumulates in reticuloendothelial system** (spleen, Kupffer cells, bone marrow, and lymph nodes). Iron also accumulates in hepatocytes and other parenchymal cells after saturation of reticuloendothelial cells.

Ans.2. Hemochromatosis vs hemosiderosis

Hemosiderosis is focal deposition of iron in the form of **hemosiderin in specific tissues** that does not cause widespread tissue damage. On the other hand, hemochromatosis is the condition of excess iron in the body which causes widespread tissue damage due to deposition of iron.

Ans.3. Bronze diabetes is a condition **where iron deposit in liver, pancreas, skin, heart which results in cirrhosis, diabetes, skin discoloration (brown), and arrhythmia* or heart failure.**

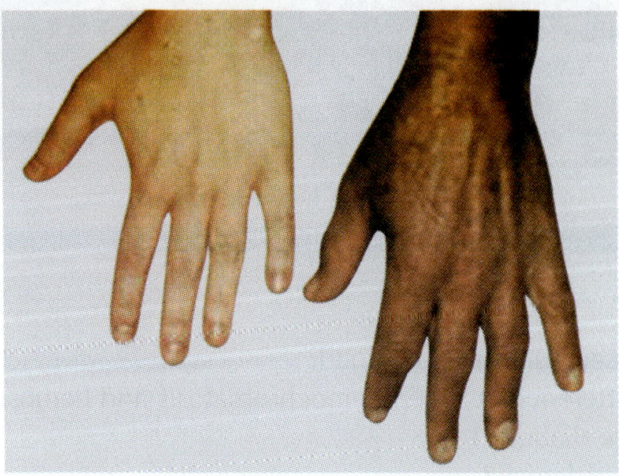

Fig. 1: Skin darkening due to deposition of iron in skin

Following are the results of various laboratory investigations:
Serum total calcium: 8.7 mg/dl (normal = 8.5–10.5 mg/dl)
Serum ionized calcium: 4.2 mg/dl (normal = 4.5–5.0 mg/dl)
Total protein: 7 g/dl (normal = 5.5–8.5 g/dl)
Serum albumin: 5.2 g/dl (normal = 3.5–5.5 g/dl)

QUESTIONS

Q.1. What may be the probable clinical diagnosis in this patient?
Q.2. Name the above two signs which the patient is showing on examination.
Q.3. What are the different forms of calcium in the blood? What is the physiologically active form of calcium?
Q.4. What is the effect of low and high plasma pH on the level of ionized calcium?

Explanations

Ans.1. The presentation in above case is typically seen in hypocalcemia.

Ans.2. The ipsilateral fascia twitching on gently stroking the side of face is known as **Chvostek sign.**

Adduction of thumb with simultaneous flexion of metacarpophalangeal joints along with extension of interphalangeal joints and flexion of wrist on inflating the BP 20 mm Hg above the systolic pressure for at least 3 minutes is known as **Trousseau's sign.**

These signs are manifested when the ionized calcium level is less than 4.7 mg/dl (normal ionized calcium level is 4.9 mg/dl). Trousseau sign and Chvostek sign are clinical tests to unmask latent tetany.

Ans.3. ***Following are different circulatory forms of calcium in the blood:***

- **Free/ionized form** (it is the physiologically active form of calcium): It is 50% of total calcium.
- **Bound form of calcium** (bound with plasma protein mainly albumin): It is 45% of total calcium.
- **Calcium complexed with citrate, phosphate and bicarbonate**: It is 5% of total calcium.

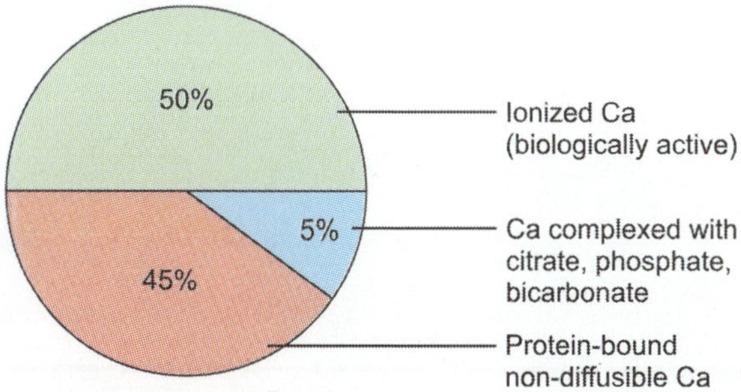

Fig. 1: Different forms of circulating calcium in plasma (50:45:5)

Total calcium which is being estimated is sum total of three different forms of calcium.

Ans.4. Plasma pH has an important impact on the level of ionized and bound calcium. At physiological pH, albumin is anion (negative charge) and calcium is cation (positive charge), hence these two bind tightly and adequately.

In acidosis when there is low plasma pH, albumin tends to be less anion and **released calcium** in free form raising the level of ionized calcium.

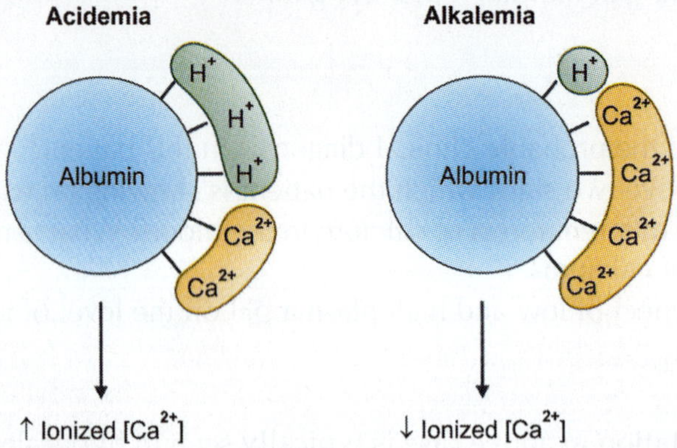

Fig. 2: Impact of acidosis and alkalosis on free and bound calcium

Reverse is also true, which means in event of **alkalosis** when pH of the plasma is high, albumin tends to be more anionic and **binds positively charged calcium tightly,** reducing the level of free/ionized calcium which may precipitate the attack of tetany (this is clinically represented in Case 67).

Bibliography

1. Higgoda R, Lokuketagoda K, Poobalasingham T, et al. Dengue fever manifesting with tetany as the first presentation of primary hypoparathyroidism: a case report. BMC Res Notes 11, 588 (2018). https://doi.org/10.1186/s13104-018-3701-2
2. Johnson MM, Patel S, Williams J. Don't Take It 'Lytely': A Case of Acute Tetany. Cureus. 2019 Oct 5;11(10):e5845. doi: 10.7759/cureus.5845. PMID: 31754580; PMCID: PMC6830852.

CASE

67

Hypocalcemia in a Case of Respiratory Alkalosis

A 23-year-old lady is admitted in emergency wards with history of rapid respiration and **involuntary spasm** in right hand (Figure below). She was in distress and was hyperventilating. She complained of tingling sensation all over the body. Mother gives the history of similar episode in past for which she was admitted, treated and was discharged in two days.

On examination, she had **twitching of small muscles** on chin and perioral areas. Respiratory rate (RR) was 38/minute. She also had carpal spasm in right hand.

Blood samples were collected immediately for arterial blood gas (ABG) analysis, for the serum electrolytes and total as well as ionized calcium.

The **ABG revealed alkalosis with a pH 7.51, pCO_2 23 mm Hg, HCO_3 21 mEq/L and low serum ionized calcium of 0.70 mmol/L.** Electrolytes were normal.

Total serum calcium level was reported as 9 mg%.

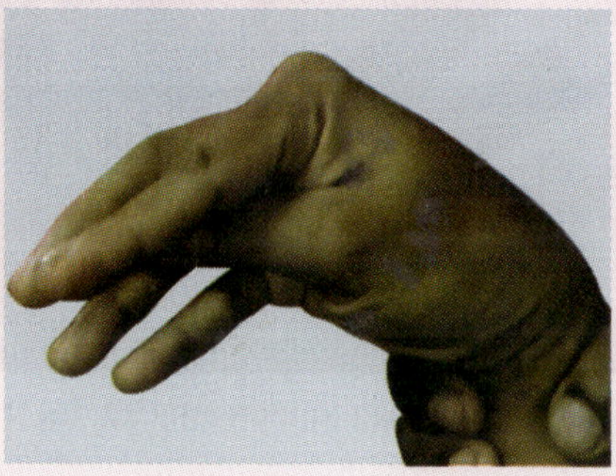

Tetanic spasm in hand (flexion of metacarpophalangeal joint with extension of interphalangeal joint along with adduction of thumb)

QUESTIONS

Q.1. What may be the reason of **tingling sensation** and spasm of hand?

Q.2. How the rapid breathing is responsible for low ionized calcium and tetany in this patient?

Q.3. Suppose you encounter such a case in your posting what will be the immediate thing you would like to do before starting the definitive treatment for this lady?

Explanations

Ans.1. Low ionized calcium (physiologically active form of calcium) is responsible for **tetanic features** in this lady.

Ans.2. Low level of ionized calcium is due to rapid breathing **(hyperventilation)** which is causing CO_2 wash off and respiratory alkalosis.

The effect of pH on the level of free and bound calcium is explained in previous case discussion.

Ans.3. The underlying problem in this patient is rapid breathing which is precipitating respiratory alkalosis. Respiratory alkalosis is the reason of tetanic spams because of low ionized calcium.

In this scenario, the rebreathing is attempted by covering the face of the patient so that patient inhales CO_2 rich air and pH of plasma is reduced balancing the level of ionized/free calcium.

Definitive treatment is 20 ml of **10% calcium gluconate** as a slow intravenous infusion over 15 minutes.

Fig. 1: Rebreathing in paper/plastic bag to increase CO_2 level in blood and lower pH

Bibliography

1. Moon HS, Lee SK, Chung JH, In CB. Hypocalcemia and hypokalemia due to hyperventilation syndrome in spinal anesthesia -A case report-. Korean J Anesthesiol. 2011 Dec;61(6):519-23. doi: 10.4097/kjae.2011.61.6.519. Epub 2011 Dec 20. PMID: 22220232; PMCID: PMC3249577.
2. Teena Bansal, Sarla Hooda, Hyperventilation causing symptomatic hypocalcaemia during labour in a parturient, Egyptian Journal of Anaesthesia, Volume 29, Issue 4, 2013, Pages 333–335, ISSN 1110-1849, https://doi.org/10.1016/j.egja.2013.05.008.

CASE

68

Dental Fluorosis

A 7-year-old boy from Rajasthan presents with **mottling* and rough ridge on teeth.** Many individuals are having similar manifestations in that village. Some of them even have **deformity of long bones.** Water of the village has high level of fluorine 2.6 mg/L (normal is 1 mg/L).

Diagnosis of dental fluorosis is made.

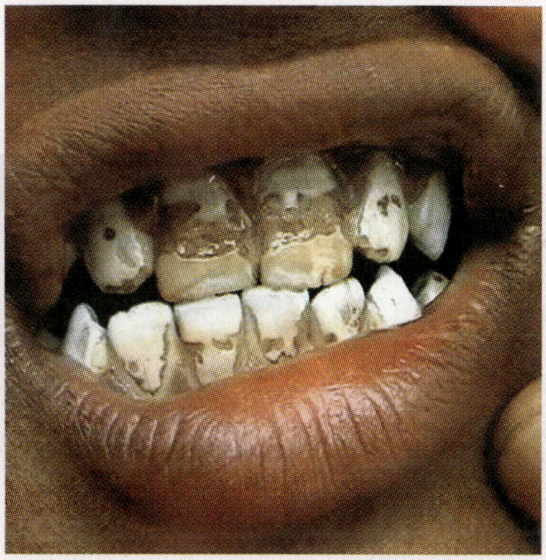

Mottling of teeth

QUESTIONS

Q.1. Comment on the diagnosis.

Q.2. What is normal RDA of fluorine?

Q.3. What are different manifestations of excess fluorine consumption?

Q.4. Why fluorine is said to be two-edge sword?

Explanations

Ans.1. This child may be suffering with **dental fluorosis** which is due to excessive consumption of fluorine in drinking water **(>5 ppm).**

Ans.2. Daily requirement of fluoride is 1–2 mg, which is easily met with 0.5–0.9 ppm level of fluorine in drinking water. Main source of fluoride is **drinking water. Sea-food and cheese** are also good sources.

Ans.3. Excess consumption of fluorine is harmful. The clinical presentation depends upon the amount and duration of excess consumption.

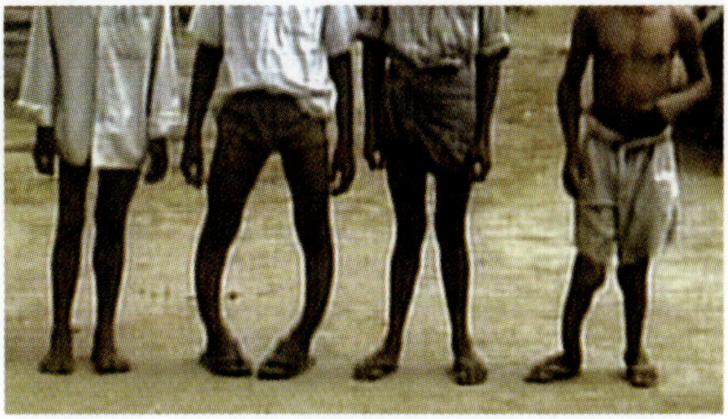

Fig. 1: Bone deformity because of excess fluorine in water in many individuals of same village

>2 ppm fluoride in water is associated with **gastrointestinal symptoms** like diarrhea, loss of appetite and loss of weight.

>5 ppm fluoride in water is associated with **dental fluorosis** which is characterized by mottled (spotting), discolored teeth with pitting as presented in above case.

If the consumption of fluorine is very **high (>20 ppm)** for chronic duration, **skeletal fluorosis** results which is characterized by calcification and ossification of ligaments, osteosclerosis and bony overgrowth resulting in deformity of bones manifesting as **kyphosis, genu valgum.**

Fluorosis is common in Rajasthan, Telangana, Punjab and in Madya Pradesh (Dhar).

Ans.4. Fluoride is a **two-edged sword:** It is said so because its excess consumption as well as deficient consumption is related with disease. **Excess consumption associated with dental and skeletal fluorosis** which has been described above.

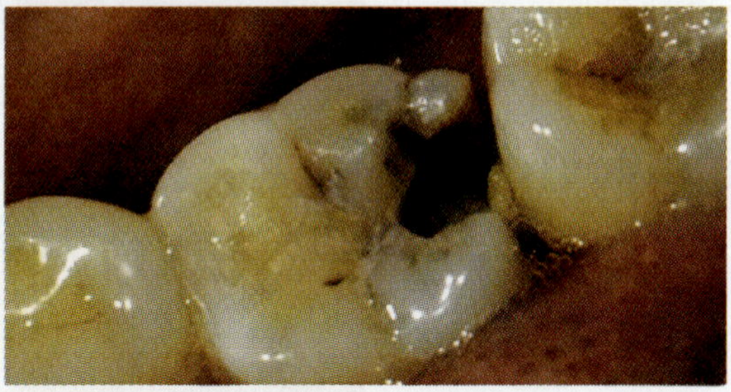

Fig. 2: Dental caries

Fluorine makes the teeth resistant to cavities by making a protective fluoroapatite layer on enamel which protects the teeth against acid attack. Fluorine makes the bone strong and makes it resistant to osteoporosis.

Deficiency (<0.5 ppm fluoride in water) is associated with **dental caries and osteoporosis.**

Bibliography

1. Khandelwal V, Nayak UA, Nayak PA, Ninawe N. Aesthetic management of dental fluorosis. BMJ Case Rep. 2013 May 22;2013:bcr2013010029. doi: 10.1136/bcr-2013-010029. PMID: 23704468; PMCID: PMC3670045.

Word Meaning

Mottling: Patches of different shades of color.

CASE

69

Keshan Disease/ Endemic Cardiomyopathy

A 48-year-old male on **total parenteral nutrition (TPN)*** presented with history of acute epigastric pain, nausea, vomiting loos of appetite and dizziness for past 2 days. Diagnosis of **dilated cardiomyopathy** was made based on ultrasonography and echo findings. Blood investigation was insignificant except **selenium level which was found to be very low 50 ng/ml (normal is 70–150 ng/ml).**

Diagnosis of dilated cardiomyopathy (DCM) associated with selenium deficiency was made.

QUESTIONS

Q.1. What is this condition known as?

Q.2. What is the role of selenium in human metabolism?

Q.3. What is the RDA of selenium?

Q.4. What is selenosis?

Explanations

Ans.1. This condition of selenium deficiency presenting with **dilated cardiomyopathy** is known as **Keshan's disease or endemic cardiomyopathy.** It is a rare disease which is mostly endemic in certain region of China, Russia and Europe, where soil is deficient in selenium content hence the crop grown in that soil tends to be deficient in this important micronutrient.

This disease has also been reported in patients who have inflammatory bowel disease, chronic malabsorption, have undergone bariatric surgery or bowel resection due to some underlying pathology (Fig. 1).

Children and nursing mother are prone to develop this disease.

Coxsackievirus B_3 infection is commonly associated in Keshan disease. In fact, occurrence of Keshan disease is multifactorial ranging from dietary, geographical and viral connections. It was first reported in rural China. There are four variants of Keshan disease: Acute, subacute, chronic and latent or compensated.

Ans.2. Selenium is required for synthesis of selenoprotein. 30 such proteins have been identified in human. Important proteins (enzymes), where **selenocysteine** is an important component, are enlisted:

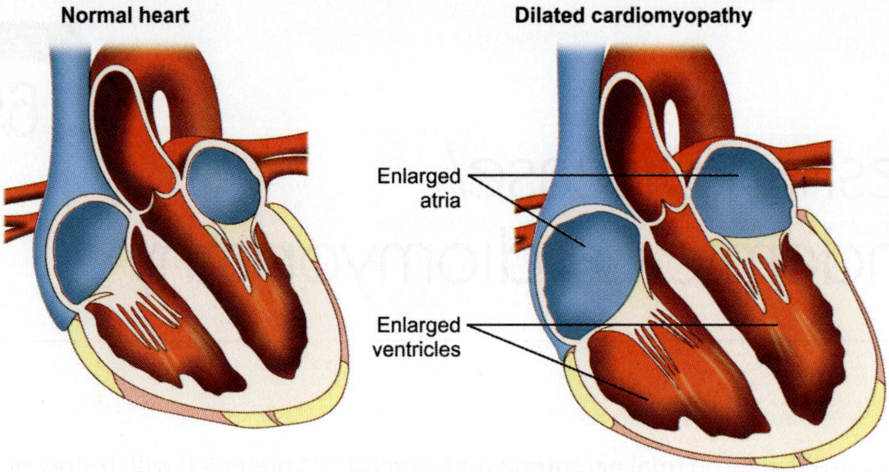

Fig. 1: Dilated cardiomyopathy

- Glutathione peroxidase
- Deiodinase
- 5′-nucleotidase
- Thioredoxin reductase
- Selenoprotein P

Glutathione peroxidase is important for antioxidant defence mechanism, hence in its deficiency increased oxidative stress results in adverse outcome.

Ans.3. RDA of selenium is **55 μg per day**. In pregnancy and lactation RDA is higher (70 μg per day).

Ans.4. **Selenosis is condition of selenium toxicity.** This is manifested as loss of hair, falling of nails, weight loss and unpleasant garlic odor in breath.

Bibliography

1. Ge K, Xue A, Bai J, et al. Keshan disease—an endemic cardiomyopathy in China. Vichows Archiv A Pathol Anat 1983;401:1–15. https://doi.org/10.1007/BF00644785
2. Munguti CM, Al Rifai M, Shaheen W. A Rare Cause of Cardiomyopathy: A Case of Selenium Deficiency Causing Severe Cardiomyopathy that Improved on Supplementation. Cureus. 2017 Aug 29;9(8):e1627. doi: 10.7759/cureus.1627. PMID: 29098137; PMCID: PMC5659335.

Word Meaning

Total parenteral nutrition: Intravenous supply of all nutrients in critically ill patients who are not able to feed themselves.

CASE

70

Indian Childhood Cirrhosis

A 4-year-old boy is coming to pediatric OPD with complain of yellow discoloration of eye and nausea vomiting for past 1 week. He was afebrile and was thoroughly investigated for the possible cause of jaundice.

Hepatitis viral markers were found absent. After ruling out all the differential diagnoses, attending pediatrician suspected a rare diagnosis of Indian childhood cirrhosis.

QUESTIONS

Q.1. What is Indian childhood cirrhosis? What are other terminologies for this disease?

Q.2. Is this disease related with excess copper consumption in copper vessels?

Q.3. How will you differentiate this disease from Wilson's disease?

Explanations

Ans.1. The infantile and childhood variety of cirrhosis was first described by Sen from Bengal in 1887. Earlier, it was known by the name **'Infantile cirrhosis' or 'Infantile biliary cirrhosis'**.

In 1960, Achar and colleagues from Chennai proposed a change in name to **'Indian childhood cirrhosis' (ICC)** denoting its affliction in young children rather than infants.

Some now suggest it to be named **idiopathic childhood cirrhosis** as the exact cause of this disease still remains obscure. It is seen in young children as progressive liver failure due to excess deposition of copper in liver revealed in liver biopsy.

Ans.2. Use of **copper and brass vessels** as a cause of this disease remains a controversial issue. Families using brass of copper utensils for milk or water storage are seen to have preponderance of such cases in children in few reports, though many authors have denied the role of copper in initiating the pathogenesis.

In a large study by ICMR, the frequency of using copper and brass vessels were insignificantly higher in cases compared to control group (55% vs 52%). Sethi et al reported no use of Cu utensils in 46% of children with ICC.

Till date, the etiopathogenesis of ICC remains a **mystery.** The exogenous toxic agent that activates the hepatocyte injury in genetically predisposed individuals is still obscure. There are no definitive clinical criteria to diagnose ICC, and the diagnosis primarily lies on characteristic histopathological features on liver biopsy.

Ans.3. Wilson's disease and Indian childhood cirrhosis (ICC) both have excess copper deposition in hepatocyte as evident on liver biopsy but in contrast to Wilson's disease, level of ceruloplasmin is high in Indian childhood cirrhosis.

Bibliography

1. https://journals.lww.com/jpgn/Abstract/1995/01000/Reversal_of_Indian_Childhood_Cirrhosis_by.6.aspx
2. Nayak NC, Chitale AR. Indian childhood cirrhosis (ICC) & ICC-like diseases: the changing scenario of facts versus notions. Indian J Med Res. 2013 Jun;137(6):1029–42. PMID: 23852284; PMCID: PMC3734708.

CASE

71

Wilson Disease

A 17-year-old male is admitted to medicine department with history of bizzare behavior, episodes of crying and screaming. On examination, he had tremors in hand, incoordinated limb movements, dysarthria and dysphagia were observed.

On visual examination, his eyes showed golden brown ring around corneal rim which was confirmed as **KF ring (Kayser-Fleischer ring)** on slit lamp examination.

Following are the results of various laboratory investigations:
Serum ceruloplasmin: 12 mg/dl (normal level is 18–35 mg/dl)
Urinary copper (24-hr collection): 270 µg (normal is 20–50 µg)
Copper content in liver (biopsy): 300 µg/g of tissue (normal level is 20–50 µg/g)

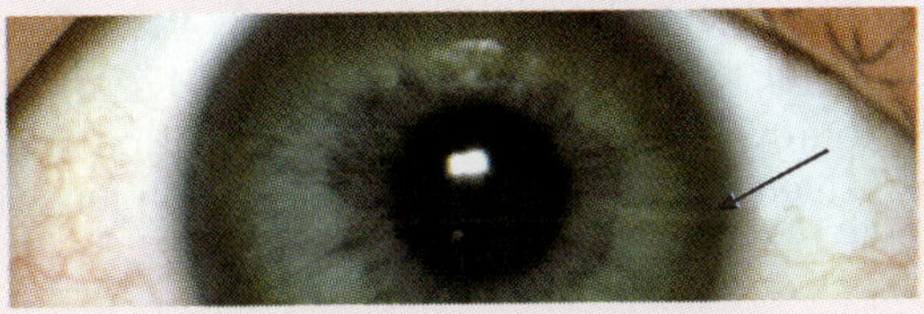

Kayser-Fleischer ring

QUESTIONS

Q.1. Identify the given disorder.

Q.2. Which micronutrient's altered metabolism is related to this kind of manifestation?

Q.3. Explain the biochemical and clinical findings in above case.

Q.4. What is the treatment suggested?

Explanations

Ans.1. This patient is suffering with **Wilson's disease** which is an autosomal recessive disorder. This disease is caused by **mutation of *ATP7B* gene** which encodes membranous **copper transporting ATPase** in the liver.

Ans.2. It is the altered metabolism of copper in Wilson's disease. **Copper transporting ATPase is the** transport protein which has got role in secretion of copper into biliary channel which helps in excretion of copper.

The defect in biliary secretion of copper results in excess accumulation of this trace element in the liver and other tissues and toxicity.

Ans.3. ***Biochemical findings:***

a. *Low ceruloplasmin level in plasma:* Level of ceruloplasmin in serum is low due to lack of incorporation of copper to apoceruloplasmin.
b. *Raised plasma and urinary copper:* As the disease advances, free copper in plasma increase which is excreted in urine and also gets deposited in brain and eye resulting in neurological and psychiatric disorder.
c. *Excess copper in liver:* Liver biopsy revealed excess copper content due to accumulation of excess copper in lack of its normal biliary excretion.

Clinical finding

Kayser-Fleischer ring: KF ring around the cornea is the **golden brown ring** which is due to deposition of excess copper in **descemet membrane of cornea.** This ring is visible by naked eye and **slit lamp examination** confirm the ring.

Psychiatric behavior: It is due to deposition of excess copper in brain.

Ans.4. ***Treatment:*** **Zinc** is the main line of treatment which not only block the intestinal absorption of copper, it also stimulates the synthesis of metallothionine in the liver which binds the copper in nontoxic form and reduced its toxicity.

Penicillamine was used earlier as **copper chelator** but presently **"Trientine" is preferred over penicillamine** because of its lesser toxicity.

Bibliography

1. Rahman S, Siddiqui NI, Paul GK, Sarker CB, Rahman KM. A case report on Wilson's disease. Mymensingh Med J. 2003 Jul;12(2):142–5. PMID: 12894051.
2. Raju K, Bangalore GN, Thuruvekere SN, Pathavanalli VN. Wilson's disease: A Clinical autopsy case report with review of literature. J Nat Sci Biol Med. 2015 Jan-Jun;6(1):248–52. doi: 10.4103/0976-9668.149210. PMID: 25810676; PMCID: PMC4367052.

CASE

72

Menke Kinky Steely Hair Syndrome

A 1-year-old male child is brought to pediatric OPD with complains of one episode of generalized fits 2 days ago and overall developmental delay. Baby is born to non-consanguineous marriage and was normal at birth.

On examination, he had fair complexion and scarce hair which was easily fragile and also limb hypotonia. Microscopic examination of hair revealed twisted hair. Blood investigation revealed following results:
Serum copper: 45 μg/dl (normal = 70–155 μg/dl)
Serum ceruloplasmin: 37 mg/dl (normal = 185–300 mg/dl)

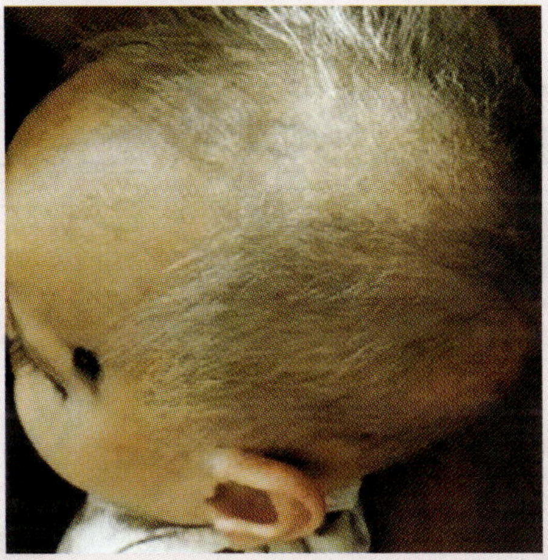

Sparse and light color hair in a child suffering with Menke disease

QUESTIONS

Q.1. What may be the possible diagnosis in this case?

Q.2. What are the enzymes get affected in this disorder? Explain their clinical association.

Q.3. Explain the hair changes which are characteristically seen in these cases.

Q.4. What is the treatment of this case?

Answers

Ans.1. This baby may be having **Menke kinky steely hair syndrome.** It is an **X-linked recessive disorder** which affects male child almost exclusively. This is due to mutation of gene of protein called **copper binding ATPase 7A** which is the transport protein for copper located at **basolateral membrane of intestinal mucosa** and is responsible for transportation of dietary copper to the blood. In mutation of this protein, the copper enters the intestinal mucosa but fails to leave it and tend to accumulate there.

This results in overall deficiency of copper to various tissues and important enzymes which require copper as their cofactor get affected with various consequences.

Ans.2. Few important enzymes which require copper and lack of their action of resulting in obvious clinical manifestations are enumerated below:

Tyrosinase: This is copper dependent enzyme. Deficiency of copper results in lack of tyrosinase activity manifesting as **hypopigmented hair** and **light color of the skin.**

Lysyl oxidase: This is copper containing enzyme and lack of action of this enzyme in this disease results in defective collagen synthesis manifesting as **vascular abnormality** and **weak hair.**

Cytochrome c oxidase: This is the enzyme of **ETC** and lack of its action results in **hypothermia.**

Ans.3. Most striking characteristic in **Menke kinky steely hair syndrome** is found in hair which is visibly **sparse, hypopigmented, short and easily fragile.** Microscopic examination of hair reveals **twisted hair (pili torti), uneven diameter of hair throughout its length (monilethrix), fracture at its shaft at multiple places (trichorrhexis nodosa).**

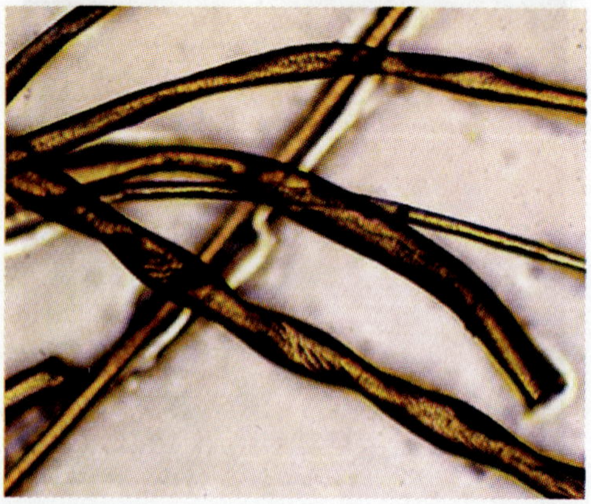

Fig. 1: Microscopic examination of hair in Menke disease [Note the twisted hair (pili torti), uneven diameter of hair throughout its length (monilethrix)]

Ans.4. Treatment of this condition **is injectable copper.**

Bibliography

1. Choudhary SV, Gadegone RW, Koley S. Menkes kinky hair disease. Indian J Dermatol 2012;57:407–9.
2. Ghosh S, Chaudhuri S. Menke's kinky hair syndrome: a case report. Dermatol Online J. 2012 Nov 15;18(11):4. PMID: 23217945.

CASE

73

Acrodermatitis Enteropathica

A 3-year-old child is brought to OPD with the history of irritability, stunted growth, diarrhea, loss of taste sensation, alopecia*, and erythematous rashes over the face, perineum, and extremities. Lesions over face and perineal area are typically symmetrical.

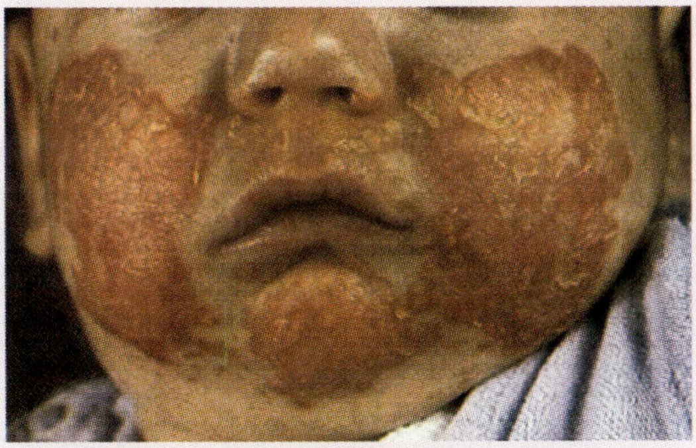

Acrodermatitis enteropathica patches

Diagnosis of acrodermatitis enteropathica is made.

QUESTIONS

Q.1. What may be the possible micronutrient deficiency in this baby?

Q.2. What is the role of zinc in human?

Q.3. What is the treatment advised?

Q.4. Why zinc is said to be two-edge sword?

Explanations

Ans.1. This baby may be suffering with the **deficiency of zinc** which is seen in **"Acrodermatitis enteropathica"** a rare **autosomal recessive disorder,** characterized by defective zinc absorption. It is characterized by diarrhea, skin lesions around mouth, nose, fingers, perineum. Lesions are typically symmetrical.

Ans.2. ***Role of zinc in body***

- Needed for storage and secretion of insulin

- Wound healing
- Spermatogenesis
- Taste sensation salivary protein **"Gusten" contains zinc** and is required for **taste sensation.** Deficiency of zinc results in **loss of taste sensation (hypogeusia).**

Ans.3. Treatment can be given by **oral supplementation of elemental zinc (60 mg twice a day).**

Ans.4. Zinc is said to be **two-edge sword,** as its deficiency as well as excess intake is associated with adverse outcome. Deficiency manifestations are discussed above with explanation of question 1.

Excess exposure to zinc is also **toxic. Toxicity of zinc** may be seen in **welding industries** where workers are exposed to welding fumes. It is characterized by fever, respiratory distress, excessive salivation, headache and sweating.

Bibliography

1. https://www.sriramachandra.edu.in/university/pdf/research/journals/jan_2007/book_14.pdf
2. Nistor N, Ciontu L, Frasinariu OE, Lupu VV, Ignat A, Streanga V. Acrodermatitis Enteropathica: A Case Report. Medicine (Baltimore). 2016 May;95(20):e3553. doi: 10.1097/MD.0000000000003553. PMID: 27196457; PMCID: PMC4902399.
3. Perafán-Riveros C, França LF, Alves AC, Sanches JA Jr. Acrodermatitis enteropathica: case report and review of the literature. Pediatr Dermatol. 2002 Sep-Oct;19(5):426–31. doi: 10.1046/j.1525-1470.2002.00200.x. PMID: 12383101.

Word Meaning

Alopecia: Loss of hair.

CASE 74

Iodine Deficiency Multinodular Goiter (Hypothyroidism)

A 26-year-old lady presents with complain of swelling in the neck for past 8 months which has been progressively increasing in size. She is giving the history of recent weight gain, lethargy and oligomenorrhea.

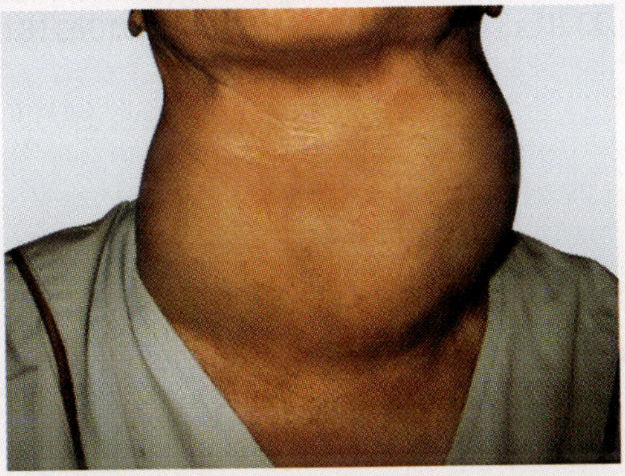

Goiter (enlargement of thyroid gland)

Many of the inhabitant of her village including children are having similar problem. Thyroid profile shows following results:

FT_3: 1.1 pg/ml (normal = 1.8–4.2 pg/ml)
FT_4: 0.40 ng/dl (normal = 0.50–1.4 ng/dl)
TSH: 8.9 mcIU/ml (normal = 0.35–5.1 mcIU/ml)

QUESTIONS

Q.1. What is the medical condition this lady is suffering with?
Q.2. What are free T3 and total T3?
Q.3. What are primary, secondary and tertiary hypothyroidisms?

Explanations

Ans.1. This lady is having **goitrous hypothyroidism** which is due to deficiency of dietary iodine. **Iodine** is required for synthesis of thyroid hormones **T_3 and T_4**. In deficiency of iodine intake, the thyroid gland enlarges due to proliferation of its follicle in an attempt to compensate thyroid hormone production and results in **multinodular goiter (MNG).**

Ans.2. The fraction of thyroid hormone which is not bound to binding proteins are known as **free T_3** and **free T_4.** These are functional forms of thyroid hormones.

Ans.3. **Disorder of thyroid gland** which is not able to produce sufficient hormone results in **primary hypothyroidism.** In order to compensate the level of TSH increases.

In **secondary hypothyroidism,** the pathology lies at the level of **pituitary gland** which is not able to produce sufficient amount of TSH.

In **tertiary hypothyroidism,** the **hypothalamus is at fault** and is not able to produce sufficient amount of TRH resulting in lowered level of TSH as well as thyroid hormone.

Bibliography

1. Booms S, Hill E, Kulhanek L, Vredeveld J, Gregg B. Iodine Deficiency and Hypothyroidism From Voluntary Diet Restrictions in the US: Case Reports. Pediatrics. 2016 Jun;137(6):e20154003. doi: 10.1542/peds.2015-4003. Epub 2016 May 10. PMID: 27244854.
2. Nyenwe EA, Dagogo-Jack S. Iodine deficiency disorders in the iodine-replete environment. Am J Med Sci. 2009 Jan;337(1):37-40. doi: 10.1097/MAJ.0b013e31817baaf0. PMID: 19155752; PMCID: PMC2634852.

CASE

75

Obesity

A 39-year-old male executive following sedentary lifestyle is seeking medical attention for his overweight. His height is 1.5 meter and weight is 82 kg. **Body mass index (BMI)*** is 36.4 kg/m^2. His waist:hip ratio is 1.5.

He is a known diabetic on regular insulin treatment for past 4 years.

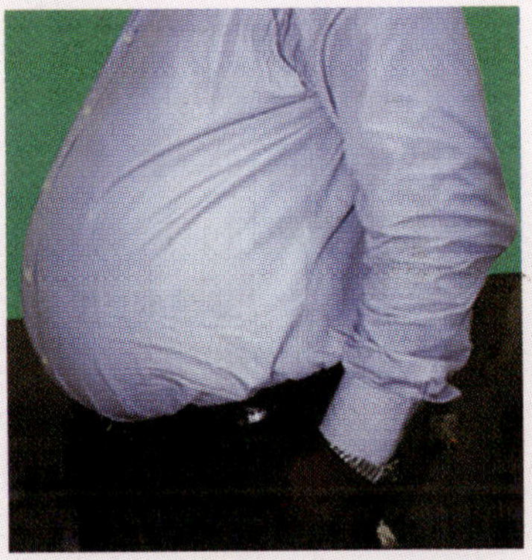

QUESTIONS

Q.1. What is obesity? Define it.

Q.2. What is the etiopathogenesis of obesity?

Q.3. Mention the role of leptin in obesity.

Q.4. What all adverse consequences a person may face because of obesity?

Q.5. What is the treatment suggested?

Explanations

Ans.1. **Obesity** is a state of **excessive amount of adipose tissue deposition.** Increased body weight is not synonym of obesity as in many cases the body weight is increased because of increased muscle mass rather than the excess deposition of adipose tissues.

BMI is most common method to assess the obesity. **BMI is defined as ratio of body weight in kg to that of height in meter square (kg/m^2).** BMI of >30 is the value taken as threshold for obesity for both men and women.

Overweight is BMI between 25 and 30.

Waist-to-hip ratio of >0.9 in women and >1.0 in men is abnormal.

This man in question is certainly obese as his **BMI is 36.4 kg/m². His waist:hip ratio is 1.5.**

Ans.2. ***Etiology of obesity***

a. *Genetic:* It is a common observation that obesity runs in family.
b. *Environment:* It plays an important role in obesity in susceptible individuals.
c. *Cultural factors:* Diet and physical activity.

Ans.3. Leptin (leptos means thin) is the peptide secreted from adipose cell. Gene of leptin is **'ob gene'**. Leptin acts through hypothalamus to regulate appetite, satiety, glucose and lipid metabolism, energy expenditure and neuroendocrine function. Leptin also acts through peripheral system to regulate immune system.

Mutation of gene of leptin/leptin receptor (ob gene/db gene) plays an important role in early onset obesity but not in common variants of obesity. Such mutations of **leptin (ob gene)** and **leptin receptor (db gene)** prevent leptin from delivering satiety signal and hence results in obesity.

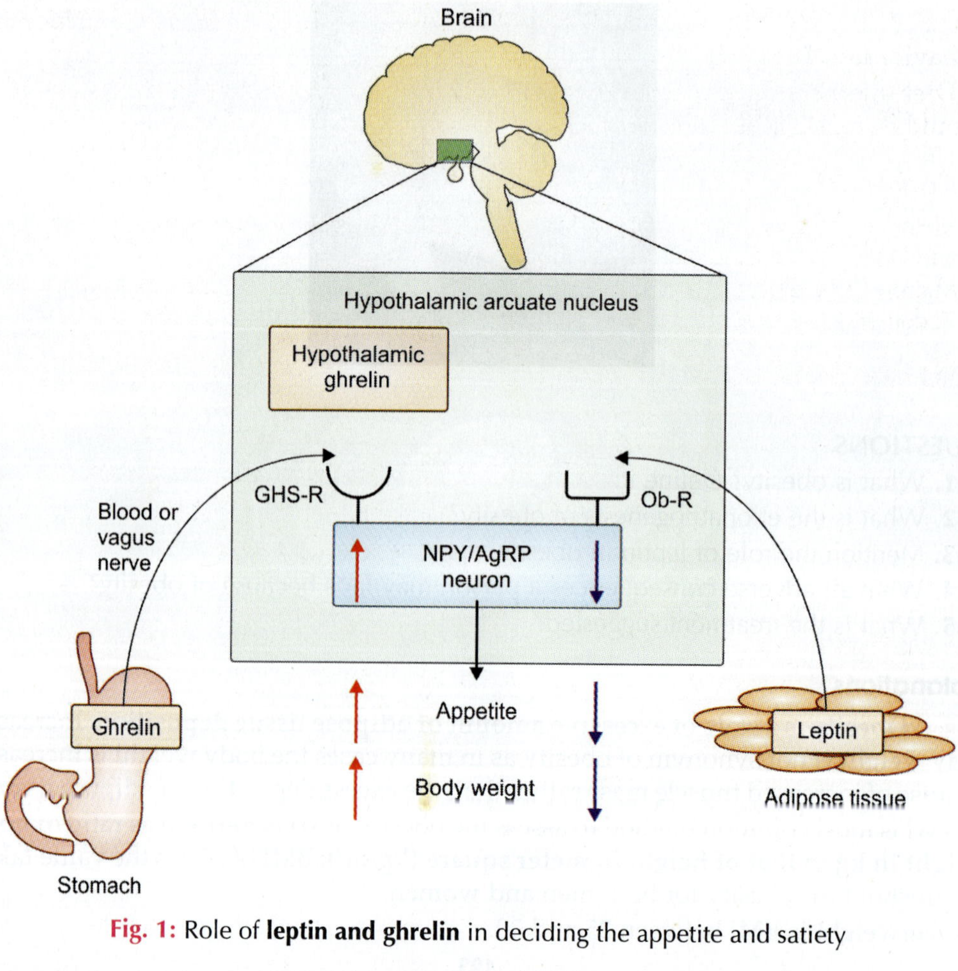

Fig. 1: Role of **leptin and ghrelin** in deciding the appetite and satiety

Ans.4. ***Adverse pathological consequences of obesity***

a. **Insulin resistance:** Obesity leads to insulin resistance (IR). Important points in this regard are:
- Raised insulin level and downregulation of insulin receptor
- Raised free fatty acid which impairs insulin action
- IL-6, TNF-α, adipokine, adiponectin secreted by adipocyte which alter insulin action.

b. **Reproductive disorder:** Male hypogonadism and female polycystic ovarian syndrome (PCOS) has been associated with obesity.

c. **Cardiovascular diseases:** Dyslipidemia with raised VLDL, LDL, TAG and low HDL is linked with obesity specially abdominal obesity.

d. **Liver disorder:** Obesity is linked with nonalcoholic fatty liver disease (NAFLD). Increased chances of cholesterol gallbladder stone.

e. **Bone:** Increased chances of osteoarthritis and gout.

f. **Cutaneous changes:** Acanthosis nigricans, i.e. darkening of skin of neck, elbow and dorsal interphalangeal space.

g. **Respiratory system:** Difficulty in breathing, obstructive sleep apnea.

Ans.5. ***Following lifestyle modifications are advised*: Diet, physical activity and behavior modification.**

Diet plan: Food items with low energy density (soup, vegetables, fruits, oat meal) should be included in diet.

Bibliography

1. Kleinendorst L, van Haelst MM, van den Akker ELT. Young girl with severe early-onset obesity and hyperphagia. *Case Reports* 2017;bcr-2017-221067.
2. Mulrow CD. Helping an obese patient make informed choices. BMJ. 1998 Jul 25;317(7153):266–7. doi: 10.1136/bmj.317.7153.266. PMID: 9677222; PMCID: PMC1113590.

Word Meaning

Body mass index (BMI): $\frac{\text{Body weight (kg)}}{\text{Height}(\text{m}^2)}$

CASE

76

Marasmus

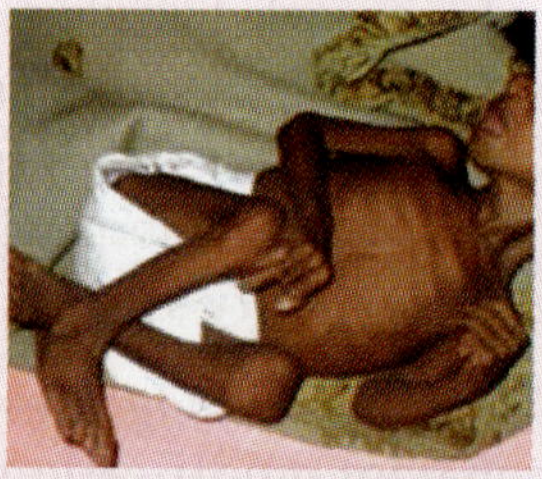

A 2-year-old child of a poor farmer is brought to the hospital with history of severe weight loss, irritability and repeated infection. He is youngest of 4 siblings. On examination, baby was **emaciated*** with dry and shrunken eye. Bony rib cage was prominent and baby had labored breathing. His body weight is 6.5 kg (normal weight of 2-year-old healthy boy is 12 kg).

Following are the results of various laboratory investigations:

Blood glucose: 55 mg/dl (normal random = 80–140 mg/dl)

Total protein: 6.5 g/dl (normal = 5.5–8.5 g/dl)

Albumin: 3.2 g/dl (normal = 3.5–5.5 g/dl)

QUESTIONS

Q.1. Comment on the condition baby is suffering from.

Q.2. How the protein energy malnutrition (PEM) is classified?

Q.3. What is the plan of management in this case?

Explanations

Ans.1. This child is suffering with **marasmus.** Marasmus is the severe form of PEM where both protein, carbohydrate, and calorie intake is considerably low. It is not uncommon to see in poor socioeconomic strata in developing countries like India. Healthy diet in adequate quantity is important for normal growth and development.

Ans.2. PEM is classified into two variants: Marasmus and kwashiorkor.

Ans.3. Adequate fluid and dietary management is essential. Food rich in both protein and carbohydrate should be provided.

Bibliography

1. Fitzroy Henry, André Briend, Vincent Fauveau, Sharon R Huttly, Mohammed Yunus, Jyotsnamoy Chakraborty, Risk Factors for Clinical Marasmus: A Case-Control Study of Bangladeshi Children, *International Journal of Epidemiology*, Volume 22, Issue 2, April 1993, Pages 278–283, https://doi.org/10.1093/ije/22.2.278

Word Meaning

Emaciated: Thin and weak body.

CASE

77

Kwashiorkor

Two brothers one aged 3 years and other aged 2 years present to OPD for complain of swelling over the body with **protruded abdomen*.** They belong to poor family and only bread winner of the family is father who is rickshaw puller. Family is mainly surviving on rice as their staple meal with no access to dal and fruits or vegetables.

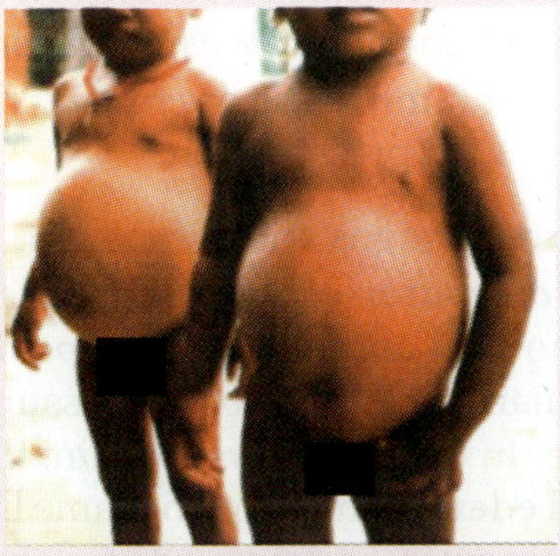

Siblings suffering with PEM (kwashiorkor): Note the protruded abdomen due to ascites

On examination, both children were **lethargic*,** abdomen was noticeably enlarged. Examination revealed **mild hepatomegaly** with **ascites*.** Lower limbs showed **pitting pedal edema*** with scaly peeling of skin at certain places. Hair examination revealed bands of **hypopigmentation.**

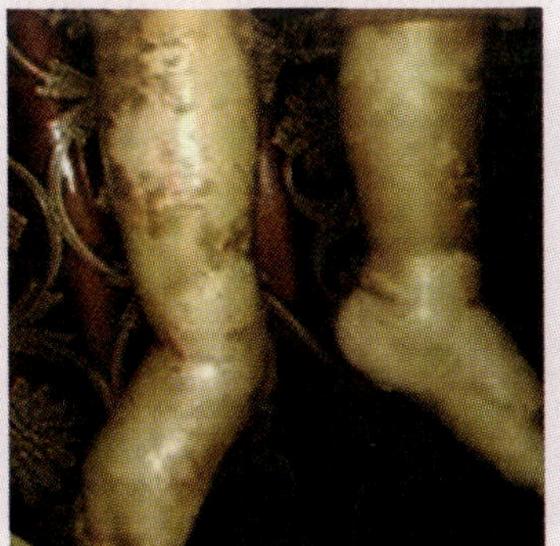

Flaky paint dermatosis in lower limb in kwashiorkor

Blood investigation revealed hypoalbuminemia in both the children.
Child A: Total protein: 6 g/dl, albumin: 3.0 g/dl (normal total protein = 5.5–8.5 g/dl)
Child B: Total protein: 5.8 g/dl, albumin: 2.7 g/dl (normal albumin = 3.5–5.5 g/dl)

QUESTIONS

Q.1. Comment on the above diagnosis.
Q.2. Explain the biochemical basis of clinical symptoms and signs.
Q.3. What is the difference between marasmus and kwashiorkor?
Q.4. What is the line of management in such cases?

Explanations

Ans.1. Both the children are suffering with "Kwashiorkor". Kwashiorkor is commonly known as **"disease of displaced child"** which is often seen in elderly sibling who is weaned off from breast milk and is put on high carbohydrate diet. It is commonly seen in children in poor family.

The **characteristic presentation** in kwashiorkor is **growth retardation, pigmentation of skin, hair, edema, protruded abdomen** because of **ascites and hepatomegaly.**

These brothers belong to poor family who are relying on carbohydrate diet without protein intake is the reason of such presentation.

Ans.2. Biochemical explanation of signs and symptoms.

a. ***Protruded abdomen due to hepatomegaly and ascites:*** Low protein intake causes decreased supply of amino acid and results in **hypoalbuminemia.** Albumin is an important protein which maintains the oncotic pressure required to hold fluid well within intravascular space. In event of low albumin, fluid oozes out of vessels and results in **ascites and pedal edema.** This is on the same line as in **nephrotic syndrome** also there is hypoalbuminemia causing edema, but in nephrotic syndrome hypoalbuminemia is because of albumin getting filtered in the urine.

   Hepatomegaly is due to fatty liver which is because of impaired lipoprotein secretion from liver which results in accumulation of excess fatty acid in the form of triacylglycerol.

b. ***Flagging in hair:*** It is characteristic sign seen in malnutrition where alternate bands of normal hair and hypopigmented hair is found in malnourished child. Period of

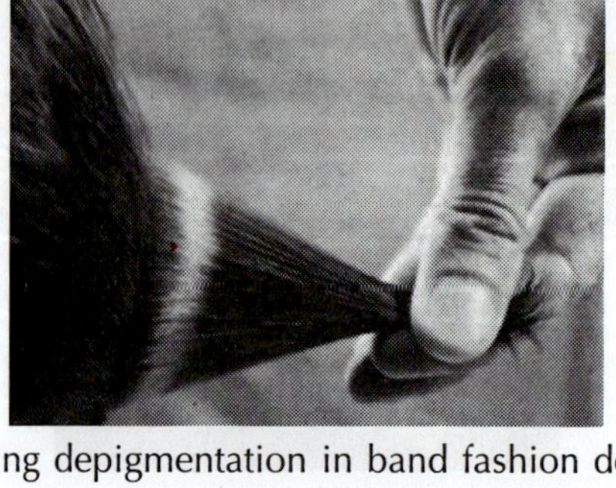

Fig. 1: Flag sign: Hair showing depigmentation in band fashion denoting the period of severe copper deficiency

severe malnutrition is reflected in **hypopigmented bands** as the lack of micronutrients specially copper in diet results in deficient action of **tyrosinase enzyme** hence no melanin synthesis.

Ans.3. Difference between marasmus and kwashiorkor: This is summarised in Table 1.

Table 1: Comparison of marasmus and kwashiorkor

Marasmus	*Kwashiorkor*
Deficient calorie and protein intake	Calorie intake is adequate, protein intake is very low
Lack of subcutaneous fat	Subcutaneous fat is reduced, not completely absent
Skin is dry	Edematous shiny skin
No flagging hair	Flagging hair
Hepatomegaly not seen	Hepatomegaly due to fatty liver seen
Hungry child (voracious appetite)	Appetite is poor
Prognosis is better	Poor prognosis

Ans.4. These children should be started on balanced diet, skin care to be given for dermatitis. For hepatomegaly liver tonics having **lipotropic agents*** need to be given.

Bibliography

1. Mori F, Serranti D, Barni S, *et al.* A kwashiorkor case due to the use of an exclusive rice milk diet to treat atopic dermatitis. *Nutr J* 14, 83 (2015). https://doi.org/10.1186/s12937-015-0071-7
2. Palm CV, Frølich JS, Snogdal LS, Støving RK. Kwashiorkor: an unexpected complication to anorexia nervosa. BMJ Case Rep. 2016 Nov 15;2016:bcr2016215638. doi: 10.1136/bcr-2016-215638. PMID: 27852657; PMCID: PMC5129140.

Word Meaning

Protruded abdomen: Distension or swelling of abdomen.

Lipotropic agents: Agents which are responsible for fatty acid mobilization from liver cell by enhancing VLDL formation and secretion. They are:

- Choline
- Lecithin
- Inositol
- Methionine
- Betain

Lethargic: Tiered and unwilling to do anything

Ascites: Fluid accumulation in peritoneal cavity causing abdominal swelling

Pitting pedal edema: Swelling in feet which leaves pressure mark/indent after the pressure is applied.

CASE 78

Lead Toxicity: Plumboporphyria and Lead-induced Anemia

A 6-year-old male child of a painter is brought to pediatric OPD by mother with history of repeated attack of abdominal pain, irritability and constipation for past 2 months. On examination, child was anemic and irritable and lacked attention.

Following are the results of various laboratory investigations:
Hb: 8 g/dl (normal = 12–15 g/dl)
Blood lactate level: 45 mg/dl (normal = <1 mg/dl)
Peripheral smear: **Microcytic hypochromic anemia** with **basophilic stippling* of RBC.**

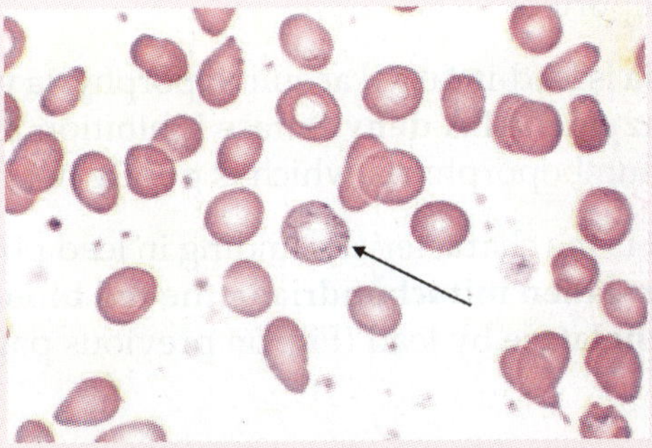

Basophilic stippling of RBC

Ring sideroblast is seen in bone marrow examination.

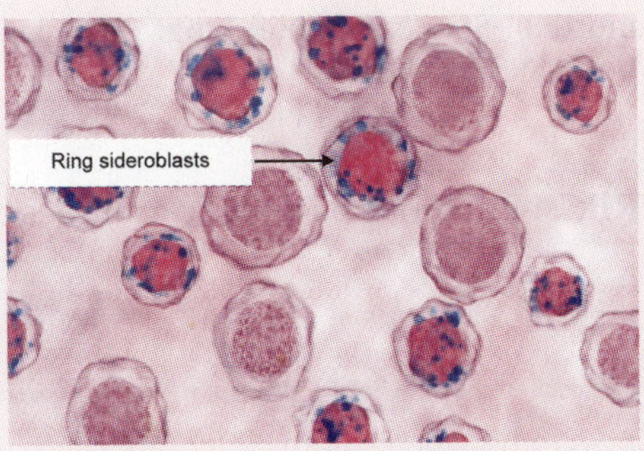

Ring sideroblast seen in RBC precursor due to deposition of iron

A diagnosis of lead poisoning is made.

QUESTIONS

Q.1. Why is the child anemic in lead poisoning?
Q.2. What are the risk factors of lead poisoning?
Q.3. What is the acceptable level of lactate in the blood?
Q.4. Which enzyme is inhibited in plumboporphyria?
Q.5. What is ring sideroblast?

Explanations

Ans.1. Lead inhibits heme synthetic pathway by inhibiting as many as eight enzymes of this pathway with major impact on **ALA dehydratase and ferrochelatase.** This results in **microcytic hypochromic anemia** in lead poisoning.

Ans.2. Risk factors for lead toxicity include the ingestion of **drinking water** having high content of lead, **lead paint** or the **dust** derived from that **paint, contaminated soil** as well as secondary exposure from occupational hazards such as workers from **paint manufacture, battery factories, metallurgy, mining industry and construction. Toys, furniture, ceramic** also contain lead.

Ans.3. Acceptable level of lead in blood is **less than 5 mg/dl.** At level above **20 mg/dl, vague abdominal pain** with **constipation, irritability** and **attention deficit** are noticed, and at level **>100 mg/dl, peripheral nerve involvement** with **encephalopathy** and even death have been reported.

Ans.4. Plumboporphyria is lead-induced acquired porphyria which is due to inhibition of **ALA dehydratase enzyme. ALA dehydratase** inhibition results in very high level of ALA which causes plumboporphyria which is purely neurological porphyria.

Ans.5. Ring sideroblast is the characteristic finding in lead poisoning and is related to the accumulation of **iron-laden mitochondria in the red blood cell precursors** due to the **inhibition of ferrochelatase** by lead (Fig. on previous page).

Bibliography

1. Alqwaifly M, Bril V, Dodig D: Acute Intermittent Porphyria: A Report of 3 Cases with Neuropathy. Case Rep Neurol 2019;11:32–36. doi: 10.1159/000496420
2. Mohan G, Madan A. Ala Dehydratase Deficiency Porphyria. [Updated 2020 Jul 8]. In: StatPearls [Internet]. Treasure Island (FL): StatPearls Publishing; 2020 Jan-. Available from: https://www.ncbi.nlm.nih.gov/books/NBK560836/

Word Meaning

Basophilic stippling of RBC: Numerous basophilic granules dispersed throughout in the cytoplasm of RBC.

CASE

79

Porphyria Cutanea Tarda (PCT)

A 45-year-old male visits dermatological clinic with complain of **red erythematous itchy lesion** over chest area and on **dorsum of hand** for past 6 months.

These lesions become prominent and itchy on exposure to sunlight. He complains of easy fatigue and had lost interest in his work.

Examination revealed **multiple red brown hyperpigmented skin lesions** on dorsum of hand with scarring of healed lesions.

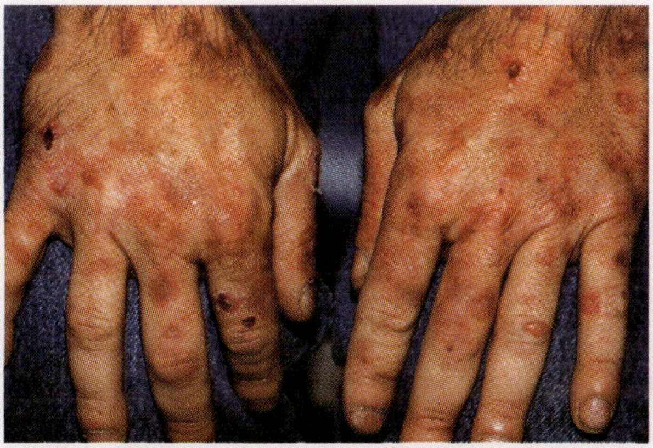

Dry itchy lesions on dorsum of hand

Following are the results of various laboratory investigations:

Ferritin level was high in blood: 1387 ng/ml (normal = <200 ng/ml)

Urine examination revealed **Soret band** and 24-hour urine porphyrin and porphobilinogen was found to be high.

Urinary porphyrin was 654 μg/24 hours (normal = ≤100)

Urine porphobilinogen was 2.45 mg/24 hours (normal = ≤1.65)

Diagnosis of porphyria cutanea tarda (PCT) is made.

QUESTIONS

Q.1. What is porphyria?

Q.2. Classify porphyria based on different criteria.

Q.3. What is the enzyme deficiency in PCT?

Q.4. What is Soret band?

Q.5. What is the treatment suggested in the above case?

Explanations

Ans.1. **Porphyria** is defined as group of disorders due to abnormalities in the pathway of biosynthesis of heme.

Ans.2. Porphyrias may be **classified as erythropoietic or hepatic on** the basis of primary site/organ of overproduction or porphyrins or its precursor.

Erythropoietic porphyria includes:
- Congenital erythropoietic porphyria (CEP)
- Erythropoietic protoporphyria (EPP)
- X-linked erythropoietic protoporphyria (XLP)

Hepatic porphyria includes:
- Acute intermittent porphyria (AIP)
- Porphyria cutanea tarda (PCT)
- Hereditary coproporphyria (HCP)
- Variegated porphyria (VP)

Based on clinical presentation, it may also be classified into pure neurological, pure cutaneous and porphyria with mixed presentation.

a. *Pure neurological*
   - Acute intermittent porphyria (AIP)
   - Plumboporphyria

b. *Pure cutaneous*
   - Congenital erythropoietic porphyria (CEP)
   - Porphyria cutanea tarda (PCT)
   - Protoporphyria

c. *Mixed*
   - Hereditary coproporphyria (HCP)
   - Variegated porphyria (VP)

Ans.3. **PCT is most common porphyria** worldwide and is **due to inhibition of enzyme uroporphyrinogen decarboxylase.**

Ans.4. ***Soret band in urine:*** Porphyrins in the urine show **sharp absorption at 405 nm wavelength light** (Fig. 1).

When a series of wavelength is passed through acidified urine in an instrument called **spectrophotometer,** there occurs peak absorption of light in **405 nm wavelength.** It is called **Soret band** and is specifically seen in porphyria which has excretion of various porphyrinogens in the urine.

Porphyrinogens get oxidized in the presence of atmospheric oxygen and are converted to porphyrins which show such absorption maxima at 405 nm wavelength.

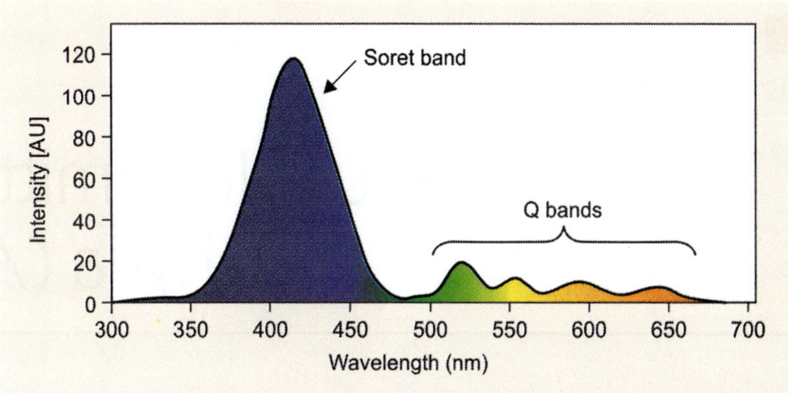

Fig. 1: Soret band in porphyria

Ans.5. Treatment in the above case is topical treatment of lesion and additionally **phlebotomy*** once in fortnight till ferritin level reaches <30 ng/ml.

Bibliography

1. Hermosilla BN, De Toro G, Molgó M. Porfiriacutáneatarda. Caso clínico [Porphyria cutanea tarda. Case report]. Rev Med Chil. 2018 Aug;146(8):943–46. Spanish. doi: 10.4067/s0034-98872018000800943. PMID: 30534876.
3. Usta Atmaca H, Akbas F. Porphyria cutanea tarda: a case report. *J Med Case Reports* 13, 17 (2019). https://doi.org/10.1186/s13256-018-1956-9

Word Meaning

Phlebotomy: Puncture of vein to withdraw blood.

CASE

80

Acute Intermittent Porphyria (AIP)

A 27-year-old lady comes with complain of **vague abdominal pain** for past 4 months. She is on oral contraceptives for past 6 months.There is no history of any cutaneous lesion. Diagnosis of neurological porphyria was suspected and urine was investigated for possible porphyrinogens. For this, **Watson and Schwartz test** was done which revealed positive result.

Diagnosis of acute intermittent porphyria (AIP) is made.

QUESTIONS

Q.1. What is the enzyme deficiency in acute intermittent porphyria (AIP)?

Q.2. How to differentiate AIP from hereditary coproporphyria (HCP) and variegated porphyria (VP)?

Q.3. What is **Watson and Schwartz test** and when it will be positive in a case of porphyria?

Q.4. Illustrate with the help of a **figure** the various steps affected in inherited and acquired porphyria.

Explanations

Ans.1. The enzyme deficiency in AIP is **PBG deaminase (UPG III synthase).** This porphyria results in **neuropsychiatric manifestation** and is characterized by lack of cutaneous manifestations.

Cutaneous manifestations in porphyria are due to **porphyrinogens** and **neuropsychiatric manifestations** are due to **raised ALA and PBG (porphobilinogens).** In acute intermittent porphyria, it is ALA and PBG which is raised giving neuropsychiatric features, level of porphyrinogens are rather normal, hence these patients lack cutaneous manifestations.

Ans.2. Hereditary coproporphyria (HCP) and variegated porphyria (VP) can be differentiated clinically from AIP by the presence of **photosensitive cutaneous lesion** and confirmed by titration of particular enzyme deficiency in erythrocytes, leukocytes and skin fibroblasts.

Ans.3. The **Watson and Schwartz** test is done to detect the presence of **porphobilinogen (PBG) in the urine** and this test is almost always positive during episodes of neuropsychiatric dysfunction. For this test to be positive, concentration of porphobilinogen in the urine should be more than three times the upper limit of normal.

Ans.4. The steps involved in various inherited and lead-induced acquired porphyria are illustrated in Fig. 1:

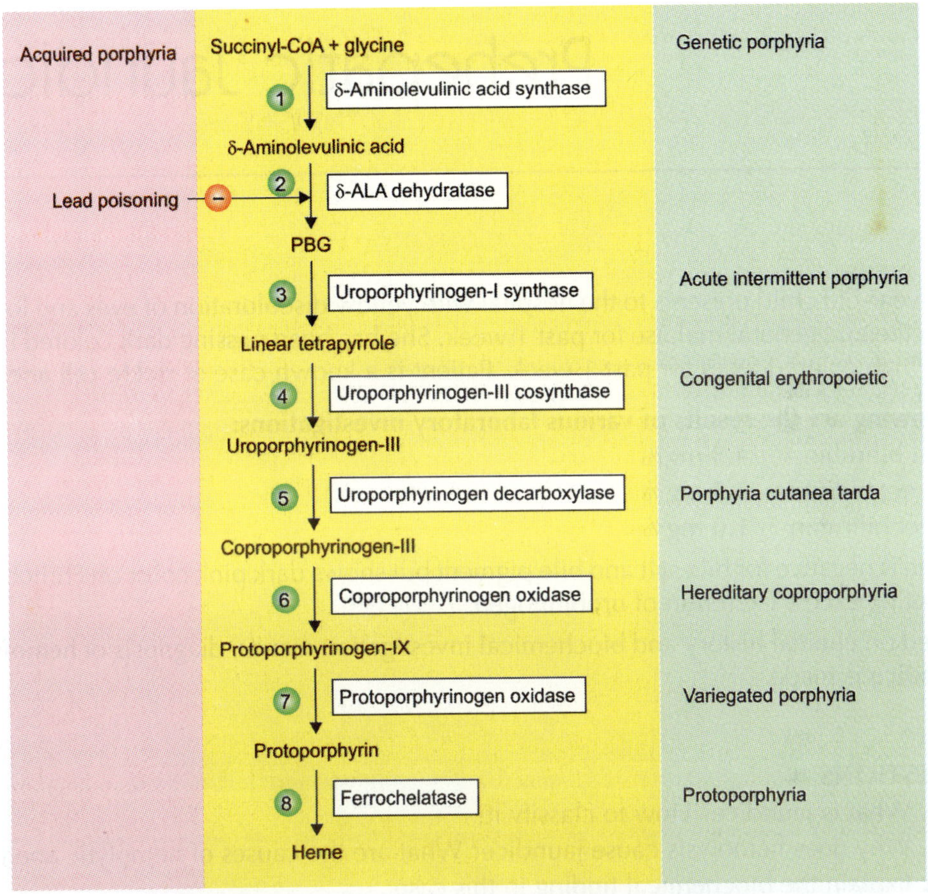

Fig. 1: Steps involved in inherited and acquired porphyria

Lead-induced porphyria is plumboporphyria which is **acquired,** rest all others are inherited porphyrias. All inherited porphyrias are inherited in an **autosomal dominant** mode except **congenital erythropoietic porphyria** which is inherited as **autosomal recessive.**

Bibliography

1. Dagens A, Gilhooley MJ. Acute intermittent porphyria leading to posterior reversible encephalopathy syndrome (PRES): a rare cause of abdominal pain and seizures. BMJ Case Rep. 2016 Jun 8;2016:bcr2016215350. doi: 10.1136/bcr-2016-215350. PMID: 27277587; PMCID: PMC4904389.
2. Ghosh S, Chaudhury PK, Goswami HK. An analysis of six cases of acute intermittent porphyria (AIP). Indian J Psychiatry. 2006 Jul;48(3):189–92. doi: 10.4103/0019-5545.31584. PMID: 20844651; PMCID: PMC2932991.

CASE

81

Prehepatic Jaundice

A 6-year-old child presents to the hospital with yellow discoloration of eyes and loss of appetite and general malaise for past 1 week. She has been passing dark colored urine and high colored feces for past 1 week. Patient is a known case of **sickle cell anemia.**

Following are the results of various laboratory investigations:

Total bilirubin	4.5 mg%
Indirect bilirubin	3.5 mg%
Direct bilirubin	1.0 mg%

Urine is negative for bile salt and bile pigment but shows dark pink color on Ehrlich test denoting excess excretion of urobilinogen.

Based on clinical history and biochemical investigation results, diagnosis of **hemolytic jaundice** is made.

QUESTIONS

Q.1. What is jaundice? How to classify it?

Q.2. Why does hemolysis cause jaundice? What are the causes of hemolytic anemia?

Q.3. Explain the biochemical finding in this case.

Q.4. How sickle cell disease is responsible for hemolytic anemia?

Explanations

Ans.1. ***Jaundice:*** Disruption in normal metabolism of bilirubin results in a condition called jaundice. Jaundice is characterized by **yellow discoloration of skin and mucous membrane** along with **pruritus*** and increased level of bilirubin in the plasma.

Normal plasma level of total bilirubin in healthy adult is less than 1.2 mg/dl. If the level of bilirubin rises beyond 1.2 mg/dl, it is known as **hyperbilirubinemia** and jaundice becomes clinically evident when the level of bilirubin in the plasma goes beyond 3 mg/dl.

Level of total bilirubin in the range of >1.2 to 3.0 mg/dl results in **latent/subclinical jaundice.**

Jaundice may be either because of overproduction, impaired metabolism or underexcretion of bilirubin in the body.

Based upon the underlying pathogenesis, jaundice is classified accordingly. Following are the **types of jaundice based on the step involved in disruption of bilirubin metabolism.**

- Prehepatic
- Hepatic
- Posthepatic

Ans.2. Hemolysis results in excessive production of heme which after its metabolism results in excess production of bilirubin. Causes of hemolysis may be genetic or acquired.

Following are conditions which result in hemolytic jaundice in clinical setting:

a. **G6PD deficiency** (glucose-6-phosphate dehydrogenase deficiency)
b. Malaria when schizonts break the RBC
c. Autoimmunity
d. Hemoglobinopathy
e. **Rh incompatibility (erythroblastosis fetalis/hemolytic disease of newborn)**

Ans.3. Hemolytic jaundice, also known as prehepatic jaundice, is due to overproduction of bilirubin which is beyond the liver capacity to handle either for conjugation or for excretion or for both.

As it can be anticipated, in this type of jaundice, plasma level of unconjugated bilirubin is high.

Hepatocyte functioning is normal and is not affected by disease process in prehepatic jaundice, in fact liver is working beyond its capacity to conjugate above normal level for bilirubin which is reaching the intestinal lumen. This results in excess production of **urobilinogen and stercobilinogen** which are getting excreted in excess amount in urine and faeces respectively in prehepatic jaundice giving high color urine and faeces.

Ans.4. In sickle cell disease, there are rupture of RBC due to repeated episode of **sickling/unsickling cycle** which results in increased fragility of RBC. For more detail please *see* Case 86.

Bibliography

1. Bensinger TA, Gillette PN. Hemolysis in Sickle Cell Disease. Arch Intern Med. 1974;133(4):624-631. doi:10.1001/archinte.1974.00320160118010
2. Srinivasan A, Gourishankar A. A Differential Approach to an Uncommon Case of Acute Anemia in a Child With Sickle Cell Disease. Glob Pediatr Health. 2019 May 7;6:2333794X19848674. doi: 10.1177/2333794X19848674. PMID: 31106251; PMCID: PMC6506923.

Word Meaning

Pruritus: Itching.

CASE

82 Rh Incompatibility (Erythroblastosis Fetalis) Presenting as Kernicterus

Rh negative mother gives birth to Rh positive child. First child escapes the insult but the second child is born with severe hemolysis and may develop kernicterus. Attending gynecologist makes the diagnosis of erythroblastosis fetalis.

QUESTIONS

Q.1. What is Rh incompatibility? Why does first child escape unaffected and major burden of hemolysis is seen only in the second child?

Q.2. What type of bilirubin will be seen in excess quantity in Rh incompatibility?

Q.3. What is kernicterus? How it is to be treated?

Q.4. What treatment is given to mother antenatally?

Explanations

Ans.1. ***Rh incompatibility (erythroblastosis fetalis/hemolytic disease of newborn)***

It is the hemolytic disease observed mostly in the **second child** when Rh negative mother conceives with Rh positive child. The Rh antigen from the first child **crosses placenta** and induces production of **Ig antibody in maternal circulation.**

Most of the time the first child escapes the devastating complications but the second child develops hemolysis intrauterine or at the time of birth when large quantity of this maternal antibody enters the fetal circulation and causes extensive damage to RBC resulting in severe hemolysis.

It requires urgent attention and **exchange transfusion*** given.

Ans.2. In hemolytic jaundice, it is predominantly **unconjugated bilirubin** which is excess in quantity.

Most of the time, the level of this unconjugated bilirubin does not exceed beyond 5 mg/dl and albumin is able to bind it making it impermeable across blood–brain barrier, but if level of unconjugated bilirubin exceeds 20 mg/dl, it outreaches the albumin capacity to bind it completely and this enters the brain crossing blood–brain barrier*.

In brain, this gets deposited resulting in irreversible brain damage. This is called kernicterus. This is a dangerous condition which is lethal many a times and if baby survives, it results in mental retardation and spasticity along with seizures.

Ans.3. *Following are the **treatment strategy** for kernicterus*

- **Phototherapy** is the mainstay of treatment in which newborn is exposed to ultraviolet light which converts water-**insoluble Z-isomer of bilirubin to water-soluble E-isomer** and promotes its secretion via urine.
- **Phenobarbiturates** to induce liver conjugating capacity.
- **Plasmapheresis** in extreme cases when the level of bilirubin **exceeds 20 mg/dl.**

Ans.4. Mother should be **immunized antenatally.**

Bibliography

1. Bi SH, Jiang LL, Dai LY, Zheng H, Zhang J, Wang LL, Wang C, Jiang Q, Liu Y, Zhang YL, Wang J, Zhu C, Liu GH, Teng RJ. Rh-incompatible hemolytic disease of the newborn in Hefei. World J Clin Cases. 2019 Oct 26;7(20):3202-3207. doi: 10.12998/wjcc.v7.i20.3202. PMID: 31667170; PMCID: PMC6819308.

Word Meaning

Exchange transfusion: Replacement of patient blood from donated blood.

Blood–brain barrier: Semipermeable microvasculature of central nervous system allowing passage of selective molecules into the brain parenchyma.

CASE

83

Hepatic Jaundice

A 23-year-old male is coming with complain of high grade fever for past 10 days which is associated with nausea and loss of appetite. He complains of yellow discoloration of eye and skin for past 1 week.

On examination, he was found to be icteric and liver was mildly enlarged and was tender.

Following are the results of various laboratory investigations:

Blood biochemistry:

Total bilirubin	5.2 mg% (normal = 0.2–1.2 mg/dl)
Indirect bilirubin	3.7 mg% (normal = 0.2–1.0 mg/dl)
Direct bilirubin	1.5 mg% (normal = 0.0–0.2 mg/dl)
SGOT	1145 IU/L (normal = 15–45 IU/L)
SGPT	692 IU/L (normal = 15–45 IU/L)
Serum ALP	220 U/L (normal = 98–298 IU/L)

Urine examination:

Bile pigments	Positive
Urobilinogen	Positive
Bile salts	Positive

HBsAg* (Australian antigen) was positive.

Attending physician makes the diagnosis of hepatic jaundice of viral etiology and further workup is planned to diagnose the cause.

QUESTIONS

Q.1. Why in hepatic jaundice both conjugated and unconjugated bilirubin tend to increase?

Q.2. What are the possible causes for hepatic jaundice?

Q.3. What is the normal level of direct, indirect and total bilirubin in the plasma?

Q.4. What is the development history of bilirubin analysis?

Q.5. Explain the principle of Malloy-Evelyn method.

Explanations

Ans.1. In hepatic jaundice, the main etiology is **hepatic dysfunction** by which hepatocytes are not able to conjugate the bilirubin and/or not able to excrete it to biliary canaliculi due to edematous cells blocking the biliary channel. This type of jaundice hence results in excess plasma level of unconjugated as well as conjugated bilirubin.

Ans.2. Most important reason of hepatic jaundice is infection. Infection by **hepatitis A, B, C, D, E viruses** are commonly encountered in day-to-day clinical practice. The resulting hepatitis not only impairs the uptake of bilirubin and its conjugation rather because of edematous cells, the blockage of **biliary canaliculi** impairs the excretion of conjugated bilirubin in intestinal lumen as stated above.

In addition to the above, physiological jaundice of newborn is also included in category of hepatic jaundice as there is underlying problem of bilirubin uptake by hepatocytes and impaired conjugation due to immature hepatocyte. This results in excess amount of unconjugated bilirubin.

Toxic insults to liver due to various chemicals and noxious agents also damage the liver and cause hepatic jaundice.

Ans.3. Normal value of **total bilirubin: 0.2 to 1.2 mg/dl**
Normal value of **conjugated bilirubin: 0.0 to 0.2 mg/dl**
Normal value of **unconjugated bilirubin: 0.2 to 1.0 mg/dl**

Ans.4. *Overview of bilirubin analysis developmental history*
Ehrlich in 1883 first described a method by which urinary bilirubin makes a colored complex with diazotized sulfanilic acid. This was classically known as **"Diazo reaction"**.

In 1913, **van den Bergh** proposed a method where diazo reaction could be applied for serum sample as well in presence of an **accelerator.**

In 1937, **Malloy-Evelyn** proposed a method where **50% methanol** was used as an accelerator for quantification of plasma bilirubin.

Followed by this in the year 1938, **Jendrassik-Grof** proposed a method in which plasma bilirubin was quantitated using **caffeine-benzoate-acetate as an accelerator.**

Today most of the methods of estimation of bilirubin is based on **Malloy-Evelyn** method, in spite of the fact the Jendrassik-Grof method is considered as candidate reference method by AACC (American Association of Clinical Chemistry).

Ans.5. *Malloy-Evelyn Method*
Principle: This method is based on the **principle of van den Bergh,** where plasma or serum bilirubin interacts with **diazotized sulfanilic acid to form azobilirubin** which is **reddish purple** compound and the color thus produced is measured colorimetrically at **560 nm.** The OD of color measured is directly proportional to amount of soluble bilirubin.

Test is done twice, first **before addition of an accelerator** and second **after addition of an accelerator. Methanol** is the accelerator used in this method.

The measurement in the first test (before addition of an accelerator) measures only conjugated bilirubin which is polar and soluble and measurement in the second test (after addition of an accelerator) measures total bilirubin (conjugated + unconjugated). This is due to conversion of unconjugated bilirubin to soluble and hence measurable form after addition of an accelerator (methanol).

Bibliography

1. Agrawal P, Chopra D, Gupta M, Santgaurav. Toxicity of Organic Solvent in a Young Painter Presenting as Transient Hepatitis: A Case Report. Indian J Occup Environ Med. 2020 Jan-Apr;24(1):33–35. doi: 10.4103/ijoem.IJOEM_61_19. Epub 2020 Mar 18. PMID: 32435113; PMCID: PMC7227743.
2. Furtado I, Valadares D, Nery FG. Acute hepatitis B virus infection and severe non-immune haemolytic anaemia: a rare relationship. BMJ Case Rep. 2017 Oct 24;2017:bcr2017221763. doi: 10.1136/bcr-2017-221763. PMID: 29066655; PMCID: PMC5665317.
3. Lohia P, Jinjuvadia R, May E. Profound jaundice in a patient with acute hepatitis C. BMJ Case Rep. 2013 Sep 12;2013:bcr2013200233. doi: 10.1136/bcr-2013-200233. PMID: 24031074; PMCID: PMC3794154.

Word Meaning

HBsAg: Also known as Australian antigen. It is surface protein of hepatitis B virus which is tested to confirm the infection by HBV (hepatitis B virus).

CASE

84

Posthepatic/ Obstructive Jaundice

A 46-year-old lady is admitted with complain of **pain abdomen on right upper quadrant,** vomiting from the last 3 days. History of **loss of appetite, weakness, malaise and nausea** for a week is present.

She gives the history of yellow discoloration of eyes and passing **high-colored urine** and **clay-colored stool.** There was no history of any fever, drug intake, alcoholism.

On physical examination, the patient was found to be pale and undernourished. Yellow discoloration of skin, sclera, and palate was observed. **Gallbladder** was palpable. Liver was **enlarged and tender.** There was no splenomegaly. The vitals were normal.

Following are the results of various laboratory investigations:

Blood biochemistry:

Total bilirubin	7.7 mg% (normal = 0.2–1.2 mg/dl)
Indirect bilirubin	1.7 mg% (normal = 0.2–1.0 mg/dl)
Direct bilirubin	6.0 mg% (normal = 0.0–0.2 mg/dl)
SGOT	145 IU/L (normal = 15–45 IU/L)
SGPT	92 IU/L (normal = 15–45 IU/L)
Serum ALP	420 U/L (normal = 98–298 IU/L)

Urine examination:

Bile pigments	Positive
Urobilinogen	Negative
Bile salts	Positive

Attending physician makes the diagnosis of **obstructive jaundice** and further workup is planned to diagnose the cause of obstructive jaundice.

QUESTIONS

Q.1. Explain bilirubin metabolism.

Q.2. What is the reason of pale-colored stool in obstructive jaundice?

Q.3. What test is done to assess urobilinogen in the urine?

Q.4. What are the various causes of obstructive jaundice?

Explanations

Ans.1. Bilirubin is the **catabolic end product** of heme. After breakdown of RBC in reticuloendothelial cell of liver, spleen and bone marrow heme is **converted to biliverdin** which is then converted to bilirubin.

Conversion of heme to biliverdin is catalyzed by **heme oxygenase** which produces **carbon monoxide as a byproduct.** This is the only reaction in human cell which produces carbon monoxide. Biliverdin is reduced by NADPH to bilirubin by enzyme biliverdin reductase.

Bilirubin thus formed is the transported to the liver bound to serum albumin. In the liver, bilirubin is taken up by hepatocyte and is conjugated to **bilirubin monoglucuronide (BMG)** and **bilirubin diglucuronide (BDG)** with the help of glucuronic acid in the presence of enzyme **uridine diphosphate (UDP) glucuronyltransferase.** BMG and BDG are water soluble and are secreted into the bile. Bile reaches duodenum where bilirubin diglucuronide is hydrolyzed to free bilirubin which is reduced to urobilinogen. Some of the urobilinogen escapes enterohepatic circulation and gets excreted in urine.

Urobilinogen is the colorless substance which is oxidized to urobilin upon excretion in urine or feces which gives normal color to urine and stool.

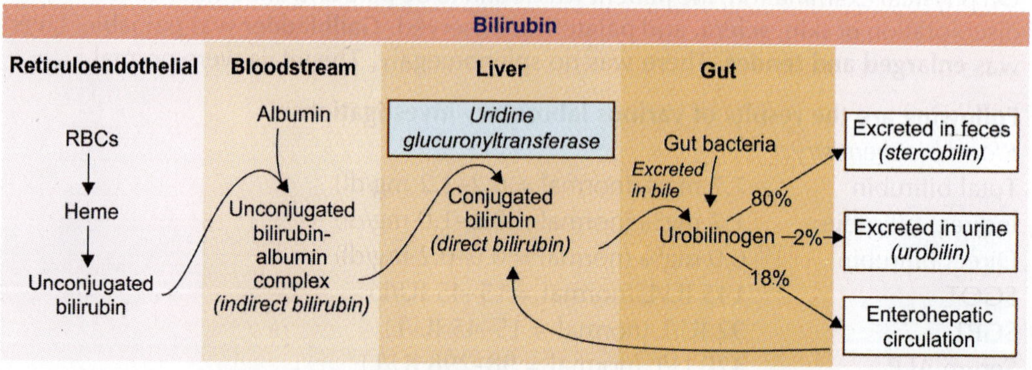

Fig. 1: Metabolism of bilirubin

Ans.2. **Urobilinogen** is synthesized in intestinal lumen from bilirubin and gets absorbed in **enterohepatic circulation** from where some amount escapes in the urine. In obstructive jaundice, bilirubin does not reach intestinal lumen, hence urobilinogen is not synthesized and in obstructive jaundice it results in clay color stool which is very characteristic finding in obstructive jaundice. **Urine also shows lack of urobilinogen.**

Ans.3. **Ehrlich test** is done to detect the presence of urobilinogen in the urine. This test is done as follows:

0.5 ml of **Ehrlich reagent** (consist of **para-dimethylaminobenzaldehyde,** HCl and distilled water) is added in 5 ml of urine and is shaken vigorously and is allowed to stand for 5 minutes. Appearance of **pink color denotes** presence of urobilinogen in the urine.

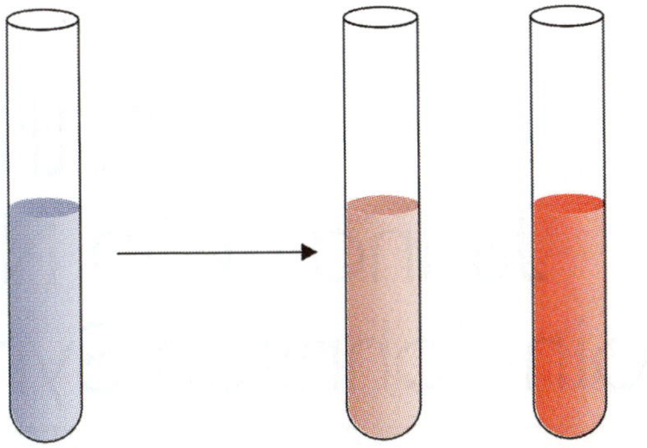

Fig. 2: Positive Ehrlich test

This is due to formation of pink color complex due to interaction of **para-dimethyl-aminobenzaldehyde** with the **urobilinogen** in acidic medium.

PBG also gives this test positive. To differentiate the UBG and PBG **Watson-Schwartz test is done with the help of chloroform and butanol.**

Ans.4. Causes of obstructive jaundice are:

- Gallbladder stone
- Malignancy head of pancreas
- Stricture of sphincter of Oddi

Bibliography

1. https://www.magonlinelibrary.com/doi/abs/10.12968/gasn.2019.17.Sup10.S22
2. Veloso N, Pires S, Godinho R. A Case of Obstructive Jaundice. GE Port J Gastroenterol. 2015 Jan 14;22(1):32–33. doi: 10.1016/j.jpge.2014.09.002. PMID: 28868367; PMCID: PMC5580134.

CASE

85 Pathological Jaundice in Newborn (Dubin-Johnson Syndrome)

A 5-year-old female child is admitted for complete check up for **recurrent episodes of jaundice since birth.** Presently she has loss of appetite, weakness, malaise and nausea for a week and is **pale and icteric.** She is passing **high-colored urine.**

She had 3 to 4 similar episodes in the past which were treated by a local physician.

She is born to a healthy nonconsanguineous parents.

On physical examination, the child was found to be undernourished. Yellow discoloration of skin, sclera, and palate was observed. The vitals were normal. Liver was enlarged and was palpable 4 cm below the costal margin.

Following are the results of various laboratory investigations:

Blood biochemistry:

Total bilirubin	5.7 mg% (normal = 0.2–1.2 mg/dl)
Indirect bilirubin	1.2 mg% (normal = 0.2–1.0 mg/dl)
Direct bilirubin	4.5 mg% (normal = 0.0–0.2 mg/dl)
SGOT	145 IU/L (normal = 15–45 IU/L)
SGPT	102 IU/L (normal = 15–45 IU/L)
Serum ALP	220 IU/L (normal = 98–298 IU/L)

Urine examination:

Bile pigments	Positive
Bile salts	Negative

Liver biopsy revealed deposition of a **dark melanin-like pigment** giving the liver a typical black cast.

Based on clinical history, biochemical investigations and histological finding in liver biopsy, the diagnosis of **Dubin-Johnson syndrome** was suspected.

QUESTIONS

Q.1. Comment on the diagnosis. What is the pathophysiology involved in this disorder?

Q.2. Explain the biochemical findings in this case. Why the urine is showing positive Fouchet's test?

Q.3. What are the other types of inherited hyperbilirubinemia?

Explanations

Ans.1. **Dubin-Johnson syndrome** is inherited in an **autosomal recessive** mode. This is characterized by **mild conjugated hyperbilirubinemia** and is due to defect in secretion of conjugated bilirubin from hepatocyte to biliary canaliculi. The protein which is defective (mutated) is **MRP-2 (multi-drug resistance associated protein 2),** which is responsible for secretion of conjugated bilirubin from hepatocytes to bile canaliculus.

Ans.2. As explained above, conjugated hyperbilirubinemia is characteristic finding in **Dubin-Johnson syndrome. Fouchet's test** detects bilirubin in the urine and the bilirubin in the urine is always conjugated one as it is water soluble. Unconjugated bilirubin is water insoluble and hence are not excreted in urine giving Fouchet's negative.

Ans.3. Number of inherited conjugated and unconjugated hyperbilirubinemia exists which is due to either defect conjugation in hepatocyte or in its excretion from the hepatocyte.

They are described below:

A. ***Disorders due to defect in conjugation of bilirubin by hepatocyte:*** These disorders are associated with unconjugated hyperbilirubinemia. Under this heading, following disorders are included:

Crigler-Najjar syndrome type I

This disease is characterized by remarkable high level of unconjugated hyperbilirubinemia to the extent of **20 to 45 mg/dl.**

This is due to severe deficiency of **bilirubin-UDP-glucuronosyltransferase** activity in the liver inherited in an **autosomal recessive** fashion.

Most of the newborns develop kernicterus and die in infancy if immediate attention is not given and vigorous phototherapy is not administered.

Crigler-Najjar syndrome type II

The underlying defect is same in CN II to that of CN I that there is deficiency of bilirubin-UDP-glucuronosyltransferase activity in the liver, but the disease severity is considerably less as the enzyme is still there to an extent of 10% of normal. This is also a genetic condition which is inherited in an autosomal recessive mode just like Crigler-Najjar syndrome type I.

Level of unconjugated bilirubin is lower than in type I (generally less than 20 mg/dl) and survival is better with fewer neurological complications.

Hepatic histology and routine liver test like AST, ALT, ALP are normal.

Type II Crigler-Najjar syndrome shows partial response to **phenobarbital therapy** while type I Crigler-Najjar syndrome shows none.

Gilbert syndrome: It is an **autosomal dominant** disorder which is due to milder deficiency of bilirubin-UDP-glucuronosyltransferase activity in the liver where one-third of enzyme activity is seen to be retained.

In this condition there is mild unconjugated hyperbilirubinemia characterized by unconjugated bilirubin **<4 mg/dl. Kernicterus** is not seen and response to phenobarbital is good. Prognosis is good.

B. ***Disorders due to defect in excretion of conjugated bilirubin by hepatocyte:*** These disorders are associated with conjugated hyperbilirubinemia. Under this heading, following two types of hyperbilirubinemia are described:
   a. Dubin-Johnson syndrome
   b. Rotor syndrome

   a. *Dubin-Johnson syndrome (DJS):* This is characterized by **mild conjugated hyperbilirubinemia** and is due to defect in secretion of conjugated bilirubin from hepatocyte to biliary canaliculi. The protein which is defective (mutated) is **MRP-2 (multi-drug resistance associated protein 2),** which is responsible for secretion of conjugated bilirubin from hepatocytes to bile canaliculus.
   This disease is inherited in an **autosomal recessive mode.**
   Characteristic dark, coarse granular pigment deposit is seen in hepatocytes in Dubin-Johnson syndrome.
   b. *Rotor syndrome:* This is a rare disorder characterized by defective uptake of conjugated bilirubin from portal circulation to the hepatocytes.
   It is seen that after conjugation in hepatocytes, majority of conjugated bilirubin are secreted in bile canaliculus by **MRP-2 (multi-drug resistance associated protein 2)** and very little quantity of conjugated proteins are secreted into portal circulation by **MRP-3.**
   The conjugated bilirubin thus secreted in portal circulation is taken up by hepatocyte with the help of yet another protein called **OATP (organic anion transport protein) 1B1 and 1B3.** In **Rotor syndrome,** there occurs mutation of OATP (organic anion transport protein) **1B1 and 1B3** which impair hepatic uptake of conjugated bilirubin by the hepatocyte, resulting in conjugated hyperbilirubinemia.

Bibliography

1. Arslan D, Avci F, Merdin A, Gunduz S, Coskun HS. Management of a patient with colon cancer and Rotor syndrome: A case report. Oncol Lett. 2014 Mar;7(3):797-798. doi: 10.3892/ol.2013.1766. Epub 2013 Dec 19. PMID: 24520296; PMCID: PMC3919936.
2. Gupta A, Tiwari P, Sachdeva P. A Case of Dubin-Johnson Syndrome in Pregnancy. Cureus. 2019 Feb 11;11(2):e4048. doi: 10.7759/cureus.4048. PMID: 31016076; PMCID: PMC6464460.
3. https://www.nejm.org/doi/full/10.1056/NEJM196005192622008
4. Jung MK, Bae MH, Kim DJ, Lee WS, Cho CM, Tak WY, Kweon YO. [A case with Rotor syndrome in hyperbilirubinemic family]. Korean J Gastroenterol. 2007 Apr;49(4):251-5. Korean. PMID: 17464171.
5. Kularatnam GA, Warawitage D, Vidanapathirana DM, et al. Dubin-Johnson syndrome and intrahepatic cholestasis of pregnancy in a Sri Lankan family: a case report. BMC Res Notes 10, 487 (2017). https://doi.org/10.1186/s13104-017-2811-6

CASE

86

Sickle Cell Disease

Five-year-old child is admitted with acute onset of pain in both the lower limb for past 4 hours. He has already been admitted twice for painful episodes of abdomen, upper limb in previous years and has been diagnosed as a case of sickle cell disease.

Presently patient is pale and febrile with body temperature 39 degree celsius. He is mild icteric as well. Spleen was palpable. He is in distress and anxious. Lower limb shows multiple healed scars of old ulcers.

Lab investigations revealed the following results:

Hb: 7.8 g%

Serum total bilirubin: 2.1 mg/dl

Direct bilirubin: 0.4 mg/dl

Indirect bilirubin: 1.7 mg/dl

Uric acid: 7.2 mg/dl (marginally elevated)

Hemoglobin electrophoresis revealed HbS band. [Hb S was 80%, Hb F: 18%, and Hb A2: 2%.]

Peripheral smear: Shows **sickle-shaped RBC**.

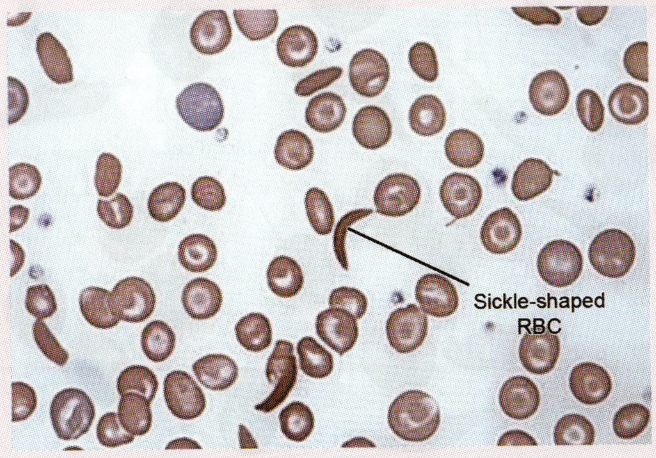

Sickle cell RBC in peripheral smear

QUESTIONS

Q.1. What is the etiopathogenesis of sickle cell disease and what do you understand by terminology 'sickle cell crisis'?

Q.2. Why RBC adopt sickle shape in deoxygenated state? Explain polymerization of HbS in deoxygenated state or why sickle cell crisis precipitate in deoxygenated state?

Q.3. Explain electrophoresis finding in sickle cell disease.

Q.4. What are the complications of sickle cell disease?

Q.5. What is the line of management in sickle cell disease?

Q.6. What is the mechanism of action of hydroxyurea in the treatment of sickle cell disease?

Explanations

Ans.1. Sickle cell disease is an autosomal recessive disease which is caused by a point mutation of the beta-globin chain gene present at chromosome 11p15.5.

This mutation changes GAG codon (meant for glutamic acid) to GTG codon (meant for valine) at the 6th position of beta-globin chain. This results in replacement of the existing glutamic acid (polar amino acid) by valine (nonpolar amino acid).

Valine at this position results in dysmorphic RBC which becomes sickle shaped which does not traverse the small capillaries and tend to occlude them. This causes sickle cell crisis characterized by sudden, severe acute pain and malfunctioning of spleen, liver, bones, joints, kidney, lung and central nervous system.

Most of the time infection is the trigger for such sickle cell crisis.

RBC tends to rupture easily giving rise to hemolytic hyperbilirubinemia. This may be the reason of pale or icteric appearance of this child.

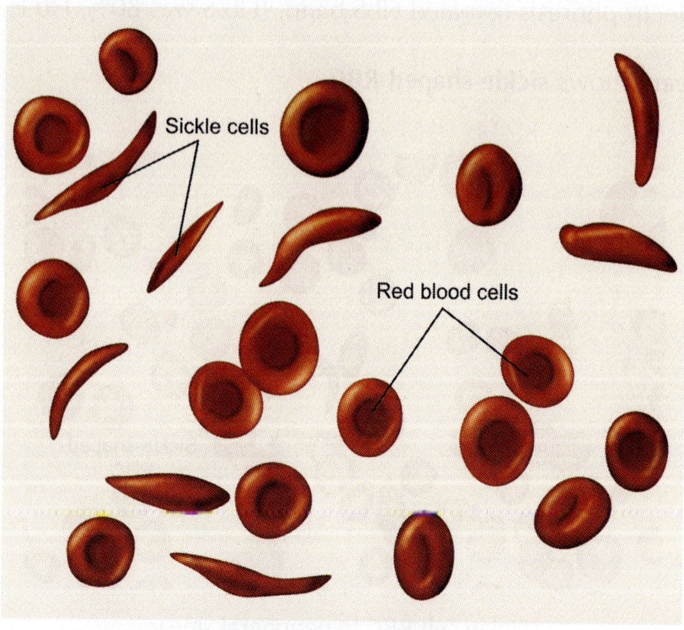

Fig. 1: Normal and sickle-shaped RBC

Ans.2. Sickling of RBC is due to polymerization of HbS in deoxygenated state. In presence of adequate amount of oxygen, the hydrophobic side chain of valine of HbS is embedded inside the globin chain and is not expressed to the surface which makes the hemoglobin soluble but in deoxygenated state the valine-6 hydrophobic side chain is exposed to the surface and precipitate the HbS in the RBC cell resulting in formation of elongated fibers.

It is important to note here that oxygen binding properties of HbA and HbS is the same, however, the solubility of HbS in deoxygenated state is reduced due to surfacing of valine-6 on the beta-globin chain in deoxygenated state.

In various research this phenomenon is described as follows:

'The replacement of glutamic acid by valine results in appearance of sticky projections in beta-globin chain, which tends to adhere with the existing depressions (sockets) in alpha chain of other hemoglobin molecule causing them to polymerize and make a thread-like structure in deoxygenated state. This kind of polymerization of HbS distorts the normal shape of RBC causing them to appear as sickle shaped (elongated and crescent shaped).' This phenomenon is depicted in Fig. 2.

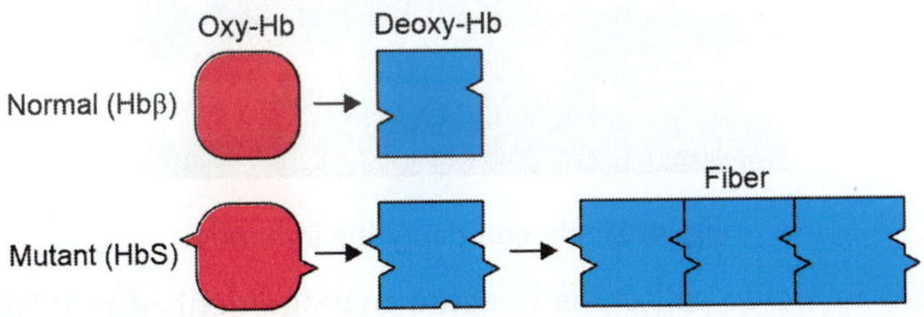

Fig. 2: Polymerization of HbS in deoxygenated state

Ans.3. The electrophoresis shows HbS band in sickle cell disease.

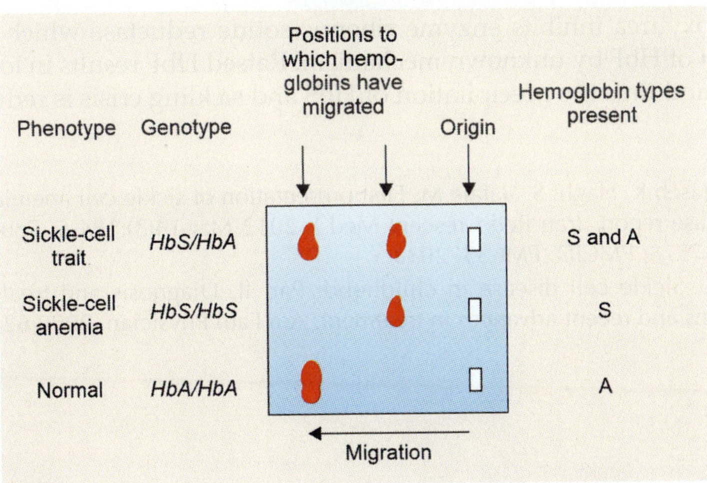

Fig. 3: Electrophoretic band pattern in sickle cell trait and sickle cell disease

Ans.4. In addition to frequent painful crisis in sickle cell disease, musculoskeletal complications are common in sickle cell disease. **Osteomyelitis** of long tubular bone is common to be seen in sickle cell disease. **Avascular necrosis of head of femur** is yet another complication affecting <5% of patients with sickle cell disease.

In addition, hand and feet may develop diffuse swelling, warmth and tenderness episodically, each episode lasting for 10 to 20 days. This is seen in children with sickle cell disease and is known as hand-foot syndrome (also known as sickle cell dactylitis).

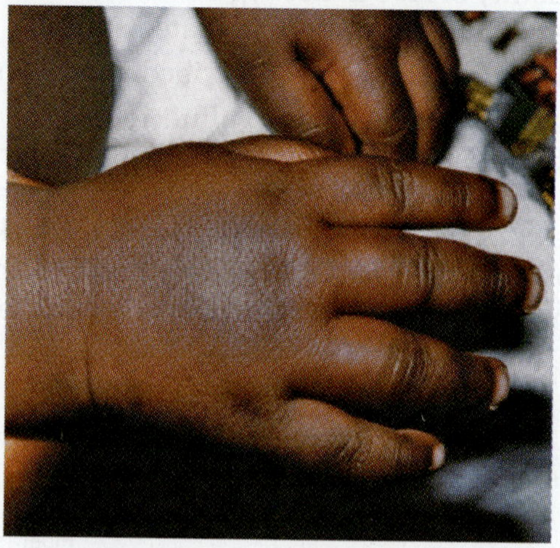

Fig. 4: Sickle cell dactylitis in hands

Ans.5. Treatment in sickle cell crisis is acute management of painful episodes and providing hydration through IV fluid along with oxygen administration. The acute episode of sickle cell crisis is often precipitated due to some febrile illness which should be adequately treated.

Hydroxyurea is the drug which is advised under closed monitoring in this disease which tends to raise the foetal haemoglobin level.

Ans.6. Hydroxyurea inhibits enzyme ribonucleotide reductase which increases the concentration of HbF by unknown mechanism. Raised HbF results in lowering of the level of HbS and thus the precipitation of HbS and sickling crisis is reduced.

Bibliography

1. Pedram M, Jaseb K, Haghi S, Vafaie M. First presentation of sickle cell anemia in a 3.5-year-old girl: a case report. Iran Red Crescent Med J. 2012 Mar;14(3):184-5. Epub 2012 Mar 1. PMID: 22737578; PMCID: PMC3372033.
2. Wethers DL. Sickle cell disease in childhood: Part II. Diagnosis and treatment of major complications and recent advances in treatment. Am Fam Physician. 2000;62:1309–14.

CASE

87

Thalassemia Major

An 8-year-old child who is a known case of **β-thalassemia major** on repeated blood transfusion is presenting with typical facial features comprising of **frontal bossing, depressed cranial vault, maxillary widening, with saddle nose.** On examination, he was found to be icteric.

Peripheral smear shows typical **hypochromic microcytic anemia** with **teardrop-shaped RBC.**

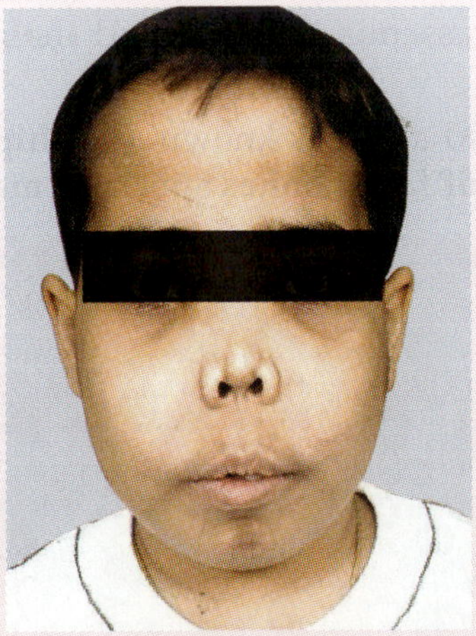

Children with typical **"Chipmunk facies" in thalassemia**

QUESTIONS

Q.1. What is the etiopathogenesis of thalassemia?

Q.2. How many clinical variants of thalassemia exist?

Q.3. What is the name given to specific facies of this child?

Explanations

Ans.1. Thalassemia is due to defect in biosynthesis of α or β-globin chain synthesis resulting in unbalanced accumulation of β and α-globin chain respectively. This is respectively called **α-thalassemia** and **β-thalassemia.**

The **mutation which causes β-thalassemia** may affect any of the following **steps** of globin chain synthesis impairing its production:

a. Transcription of gene
b. Post-transcriptional modification of mRNA
c. Translation
d. Post-translational modification

In most cases, it is either the defect in splicing of mRNA or premature termination of translation.

Ans.2. β-thalassemia may be **homozygous or heterozygous.**

There are **three clinical variants** of homozygous β-thalassemia based on severity of presentation:

a. **Thalassemia major**
b. **Thalassemia intermedia**
c. **Thalassemia minor**

β-thalassemia minor is mild and usually asymptomatic, the only manifestation is hypochromic, microcytic anemia. β-thalassemia major patients are born healthy but they become symptomatic with anemia, hepatosplenomegaly, growth retardation, jaundice and bone changes which usually develop within the first year of life, thus making regular blood transfusion and iron chelation therapy necessary for survival.

Thalassemia intermedia is a condition intermediate between the minor and the major forms. **Heterozygous** β-thalassemia is also called **thalassemia trait** and is usually asymptomatic.

Ans.3. Typical facies as stated above is known as **"Chipmunk facies"** and is due to **expansion of bone marrow of facial bones specially maxillary and frontal bones.**

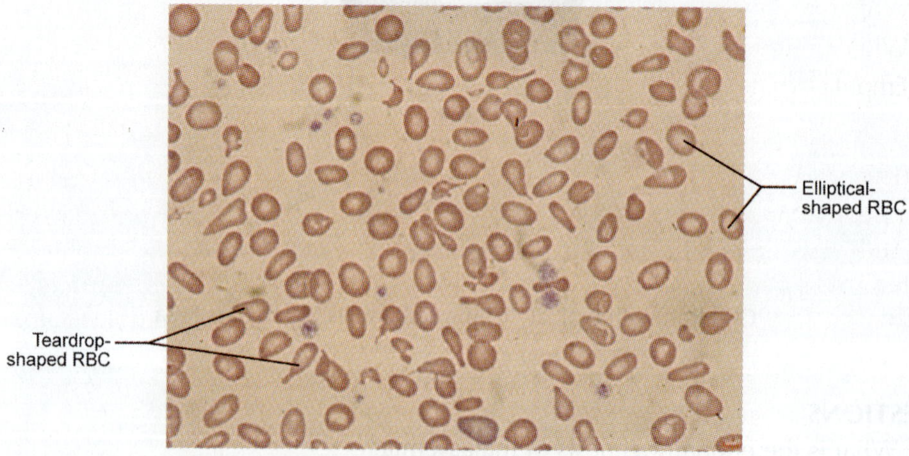

Fig. 1: Peripheral smear in a patient with thalassemia major microcytic hypochromic picture with many elliptical and teardrop RBCs

Bibliography

1. Roshan Colah, Khushnooma Italia, Ajit Gorakshakar, Burden of thalassemia in India: The road map for control, Pediatric Hematology Oncology Journal, 2017;2(4):79–84, ISSN 2468-1245.
2. Sahin C, Basaran O, Topal Y, Akin F. A thalassaemia intermedia case with concomitant left atrial thrombus. BMJ Case Rep. 2014 Jun 20;2014:bcr2013203106. doi: 10.1136/bcr-2013-203106. PMID: 24951612; PMCID: PMC4069792.

CASE

88

Marfan Syndrome (MFS)

A 21-year-old male presents to medicine OPD with complain of vague chest pain on and off lasting for 5 minutes for past 2 months. He is recently finding difficulty in vision as well. On examination, patient was **tall stature** with thin and **long extremities.**

Slit lamp examination of eye revealed **upward dislocation of lens in left eye.** He also had **cardiac involvement in form of aortic regurgitation.**

Patient was suspected to have **Marfan syndrome and was worked up.**

QUESTIONS

Q.1. What is the molecular defect in Marfan syndrome?

Q.2. What are the systems commonly involved in Marfan syndrome?

Q.3. Emphasize the role of fibrillin glycoprotein.

Explanations

Ans.1. **Marfan syndrome** (MFS) is due to **mutation of gene (FBN1)** which encodes the connective tissue protein **fibrillin 1.** It is an **autosomal dominant** condition with a reported incidence of 1 in 3000 to 5000 individuals making it one of the most common inherited disorders.

Ans.2. **Following systems are commonly involved in Marfan syndrome:**

- Ocular
- Cardiovascular
- Musculoskeletal.

Upward subluxation of lens associated with cataract, thin tall extremities will long slender fingers, high-arched palates are characteristic finding. Thumb sign is positive when on making the fist, thumb protrudes out. **Wrist sign** is positive when on holding the wrist with other hand, thumb overlaps the fingers.

Chest deformity is also associated in form of **pectus carinatum.** Cardiovascular involvement is also seen and patient may have **aortic regurgitation** and **mitral valve prolapse.**

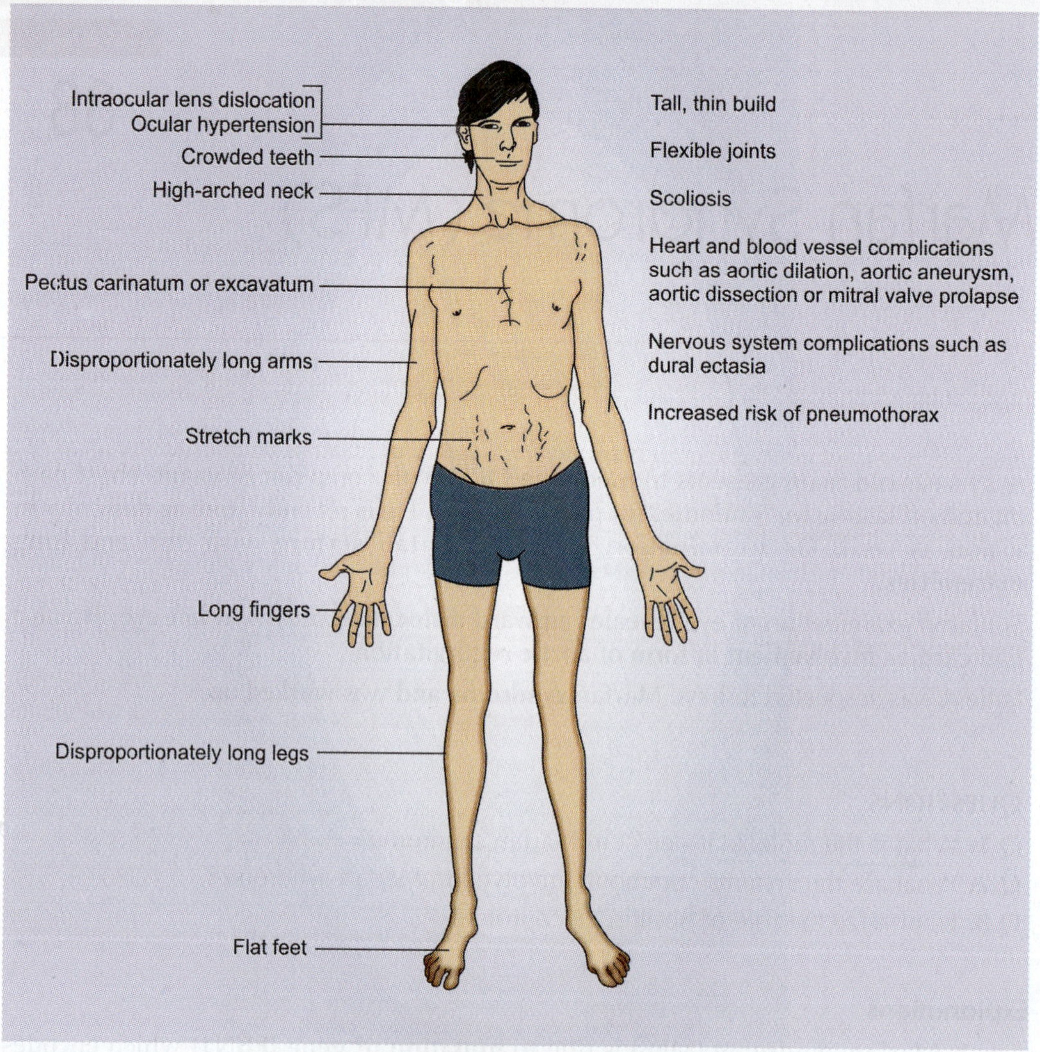

Fig. 1: Characteristic features of Marfan syndrome

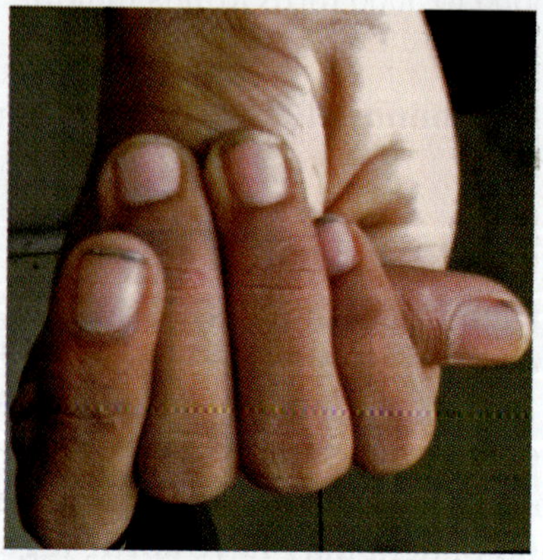

Fig. 2: Thumb sign (notice protruding thumb on making fist)

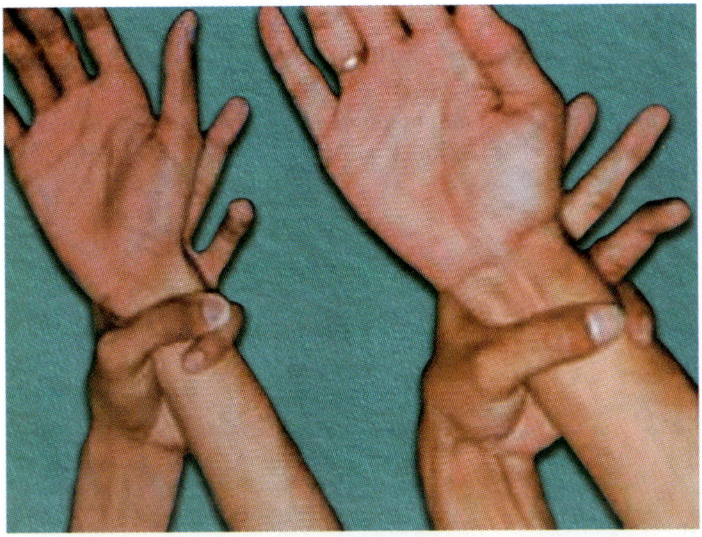

Fig. 3: Wrist sign (notice overlapping of thumb over finger(s) on holding the wrist)

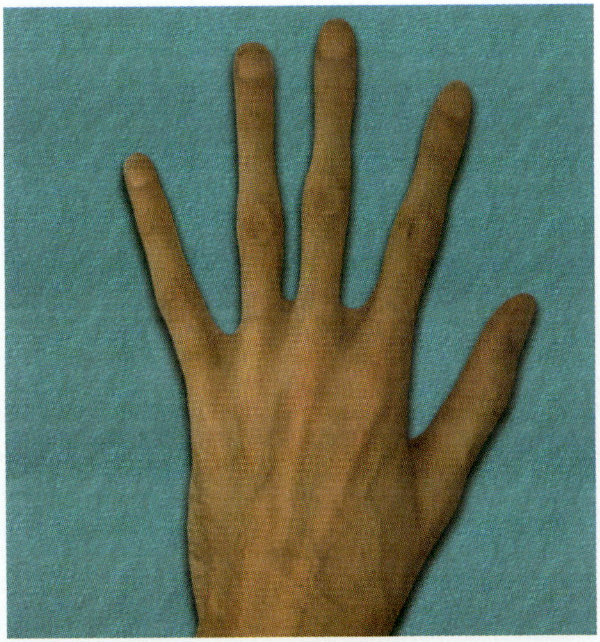

Fig. 4: Thin and long slender fingers **(arachnodactyly)** in a patient with Marfan syndrome

Ans.3. **Fibrillin 1** is a glycoprotein which is an important component of extracellular matrix of connective tissue. These **glycoproteins** interact with other component in extracellular matrix and play very important role in making elastic and nonelastic fibers in extracellular matrix, thus have important role in tissue development, repair and homeostasis.

In addition to their structural role, these fibrillins also play an important role in growth factor signaling and immune response.

Bibliography

1. Jain E, Pandey RK. Marfan syndrome. BMJ Case Rep. 2013 Dec 11;2013:bcr2013201632. doi: 10.1136/bcr-2013-201632. PMID: 24336584; PMCID: PMC3863051.
2. Naidoo P, Ranjith N, Zikalala Z, Mahoney S, Ho K. Marfan syndrome: a case report and pictorial essay. Pan Afr Med J. 2018 Jun 25;30:171. doi: 10.11604/pamj.2018.30.171.14363. PMID: 30455800; PMCID: PMC6235474.

CASE

89

Osteogenesis Imperfecta (Brittle Bone Diseases)

An 18-year-old male comes to orthopedic OPD with complain of repeated fracture of various bones of arms and foot for past 2 years. Slight trauma or even overexertion precipitates the fracture of the bones.

Radiographic examination revealed multiple healed fractures of radius, ulna, and tibia. One significant observation was **blue sclera** and **defective dentition.**

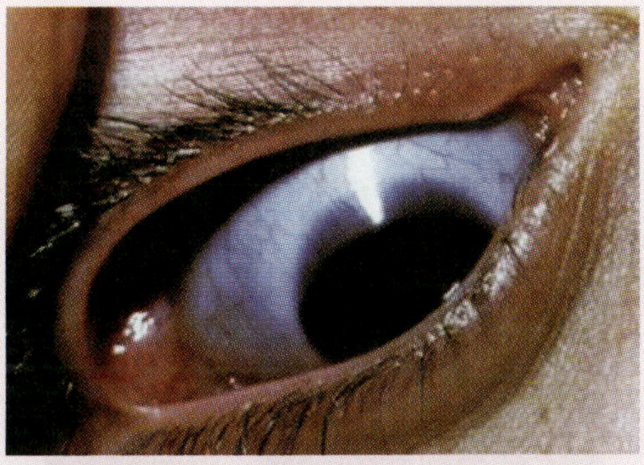

Blue sclera

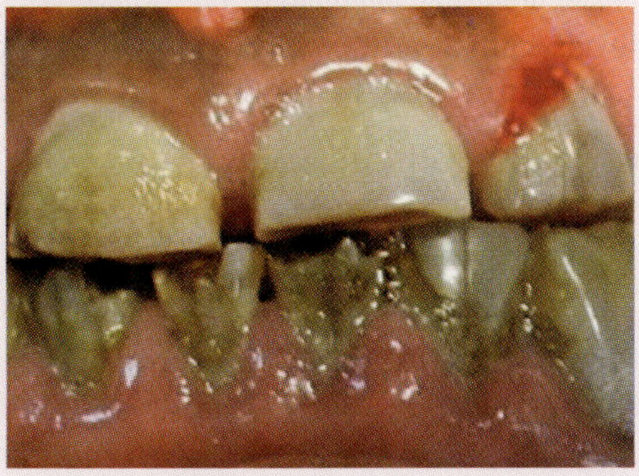

Defective dentition

Diagnosis of **osteogenesis imperfecta** is made.

QUESTIONS

Q.1. How many types of collagen are found in human? Write the distribution of important type of collagen.

Q.2. Which type of collagen is affected in osteogenesis imperfecta?

Q.3. What other type of collagen disorder do you know?

Explanations

Ans.1. So far, **28 types of collagen** have been identified and described.

The 5 most common types and their distribution are:

- *Collagen I:* Found in skin, tendon, vascular ligature, organs, bone
- *Collagen II:* Found in cartilage
- *Collagen III:* Found in reticular fibers
- *Collagen IV:* Forms basal lamina
- *Collagen V:* Found in cell surfaces, hair and placenta.

Ans.2. Osteogenesis imperfecta is caused by mutations in the **collagen type I, α_1 and α_2 genes,** which encode the α_1 and the α_2 chain of type I procollagen respectively.

This disease is characterized by repeated episodes of fracture along with **blue sclera, otosclerosis** with **hearing loss, high-arched palate, hyperlaxity of ligaments and skin.**

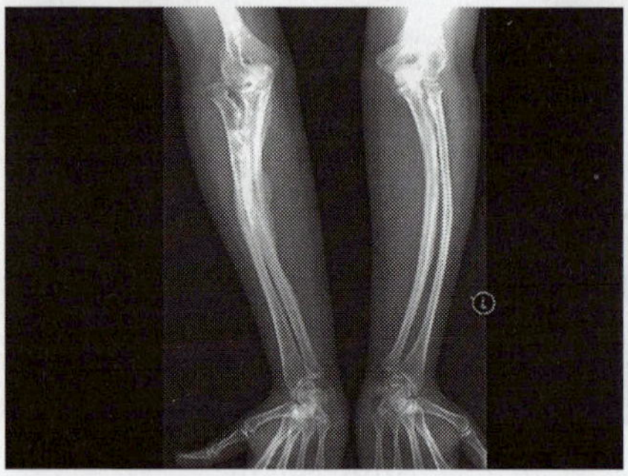

Fig. 1: Multiple fractures of long bones and deformity

Diagnosis can be made clinically. Radiographic support and confirmation by collagen analysis of **skin fibroblast** culture may be necessary in some cases:

Ans.3. Other diseases associated with collagen:

- Ehlers-Danlos syndrome
- Alport syndrome
- Chondrodysplasias
- Osteoporosis
- Knobloch syndrome
- Scurvy

Bibliography

1. Ren J, Xu X, Jian X, Wang J. Osteogenesis imperfecta type I: A case report. Exp Ther Med. 2014 Jun;7(6):1535–1538. doi: 10.3892/etm.2014.1653. Epub 2014 Mar 31. PMID: 24926339; PMCID: PMC4043568.

CASE

90

Ehlers-Danlos Syndrome

A 14-year-old female presents with complain of vague body pain for past one year and easy tiredness. She has difficulty in chewing food because of fallen teeth. Her skin was showing **elasticity and joints were hypermobile.** Eyes showed **blue sclera.**

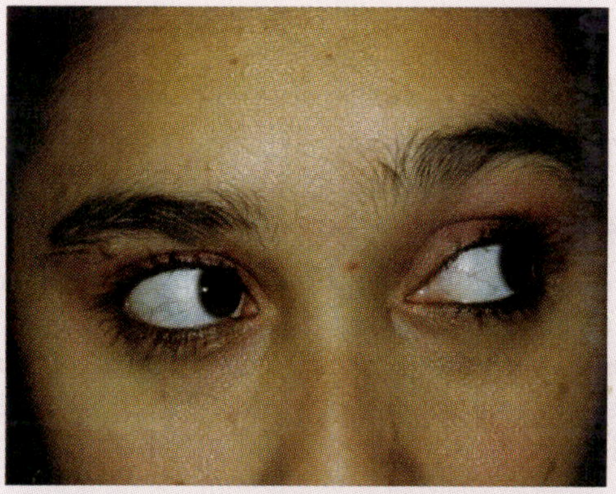

Blue sclera

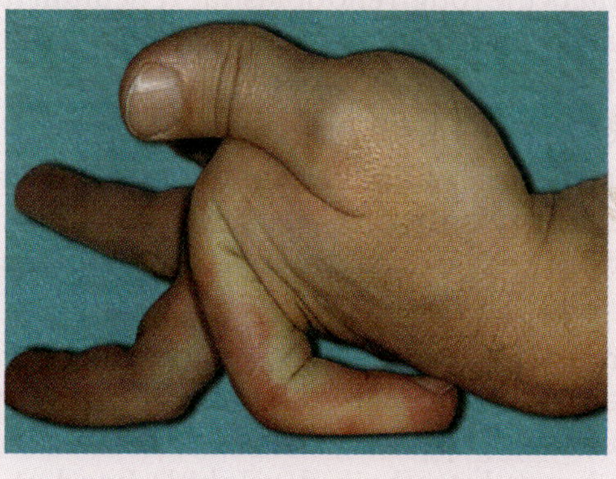

Hyperextensibility of fingers

Diagnosis of **Ehlers-Danlos syndrome** is suspected.

QUESTIONS

Q.1. Comment on the diagnosis.

Q.2. Explain collagen synthesis and post-translational modification.

Explanations

Ans.1. **Ehlers-Danlos syndrome** is a connective tissue disorder characterized by:

a. Hypermobility of the joints.

b. Hyperelasticity, fragility and softness of the skin.

c. Deficient healing of wounds.

d. Ecchymosis caused by minor traumas.

According to an estimate, more than half of the patients having **classical Ehlers-Danlos syndrome** harbor mutations in the ***COL5A1* and *COL5A2* genes,** encoding the α_1 and the α_2 chain of type V collagen, respectively. It is inherited as an autosomal dominant trait, though autosomal recessive and **X-linked recessive varieties** are also known.

Ans.2. ***Synthesis of collagen and its post-translational modification***

Collagen is synthesized by polyribosome on rough endoplasmic reticulum (RER) as **'preprocollagen'** and is extensively **modified post-translationally both intra- and extracellular locations.**

Like many other proteins synthesized on rough endoplasmic reticulum, collagen also contains **'leader' or 'signal' sequence at the N terminal end.** In addition, pre-pro-collagen has **extension peptide** at both **N and C terminal ends.** This leader/signal sequence facilitates the binding of the ribosome to the RER and directs the passage of the nascent **preprocollagen** chain to the lumen of the RER.

Post-translational modification of collagen in intracellular location

Once preprocollagen enters the rough endoplasmic reticulum lumen, the leader/signal sequence is enzymatically removed. After removal of the leader/signal sequence, the collagen chain is now known as **procollagen**.

In the lumen of RER, some of proline and lysine residues of these procollagens are converted to **hydroxyproline and hydroxylysine** by **prolyl and lysyl hydroxylase** enzymes respectively. Some of the hydroxylysine residues thus produced are modified further by glycosylation with either **glucose or galactose** residue via glucosyl transferase enzyme.

Both 'N' and 'C' terminal **extension peptides** of procollagen are rich in **cysteine** residues. 'N' terminal extension peptide cysteine residues form **intrachain** disulphide bond. 'C' terminal extension peptide cysteine residue forms **intrachain as well as interchain** disulphide bond. Interchain disulphide bond at 'C' terminal assists in formation of triple helix in the intracellular location itself.

Procollagen having interchain disulphide bond at 'C' terminal now is **translocated to Golgi apparatus,** where they are packaged in the secretory vesicles. These vesicles fuse with the cell membrane and the procollagen is released extracellularly.

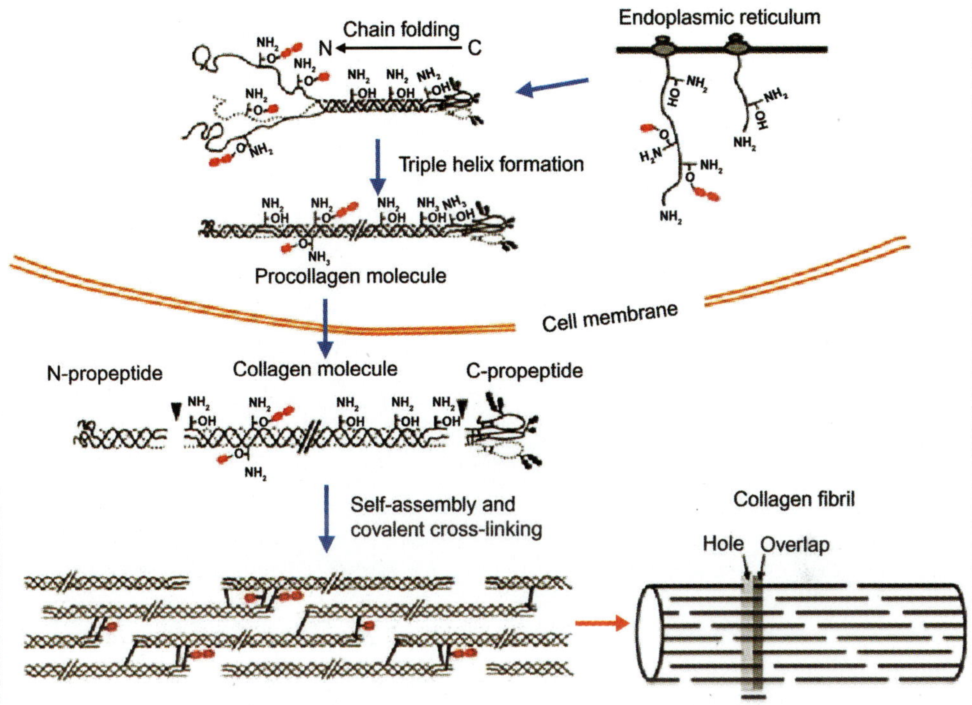

Fig. 1: Collagen synthesis and post-translational modification

Post-translational modification of collagen in extracellular location

In extracellular location, first of all the extension peptide of both 'N' and 'C' terminals is removed by **procollagen peptidases** (procollagen **aminopeptidase** and procollagen **carboxypeptidase)** releasing triple helical collagen molecule.

These triple helical collagen molecules are now assembled in the form of **collagen fibril in 'quarter staggered' alignment.** Next to this collagen fibril is arranged, collagen fibers are further stabilized by the formation of covalent **cross-links,** both within and between the triple helical units.

These cross-links are formed through the action of copper dependent extracellular enzyme **lysyl oxidase.** This enzyme oxidatively deaminates the amino groups of certain lysine and hydroxylysine residues, yielding reactive aldehydes (al-lysine and hydroxy-al-lysine).

Such aldehydes can form **aldol condensation** products with other lysine- or hydroxylysine-derived aldehydes. Aldehyde derivative of lysine and hydroxylysine may also form **Schiff bases** with the amino groups of unoxidized lysine or hydroxylysines. These reactions result in the stable covalent interchain and intrachain cross-links that are important for providing the **tensile strength** of the collagen fiber.

Bibliography

1. Kaurani P, Marwah N, Kaurani M, Padiyar N. Ehlers-Danlos syndrome—a case report. J Clin Diagn Res. 2014 Mar;8(3):256-8. doi: 10.7860/JCDR/2014/4569.4178. Epub 2014 Mar 15. PMID: 24783151; PMCID: PMC4003659.
2. Malfait F, Wenstrup R & De Paepe A. Clinical and genetic aspects of Ehlers-Danlos syndrome, classic type. *Genet Med* 12, 597–605 (2010).

CASE

91

Gout

A 55-year-old male presents with the history of **pain in the right big toe** since morning. He is not a regular alcoholic but gives the history of **consumption of alcohol** on previous night in the party along with nonvegetarian items.

On examination, the affected area was found to be swollen, tender and warm. Patient had similar episode of pain 1 year back when pain started in knee joint and then involved **big toe in the right limb.** Patient had consulted general physician who prescribed certain medications **(allopurinol)** and patient improved in 4 days.

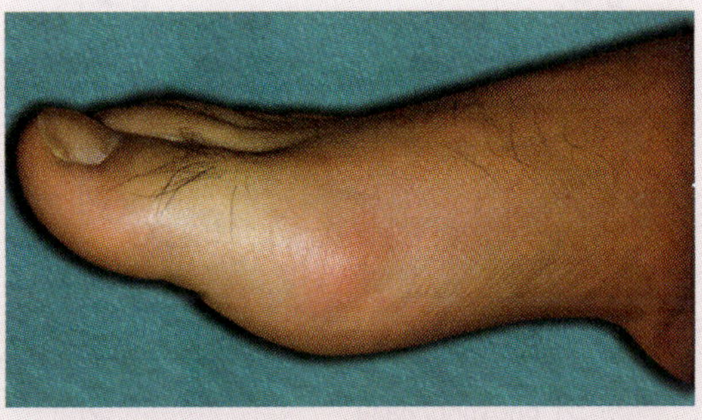

Swollen tender big toe

Biochemical investigation revealed high level of **serum uric acid which was 8.6 mg/dl.** Rest other routine parameters of KFT, LFT, lipid profile and serum electrolytes were normal.

QUESTIONS

Q.1. What may be the possible diagnosis in this case?

Q.2. Which metabolic pathway is involved in this disease?

Q.3. What is the rationale behind consumption of alcohol and precipitation of the symptoms?

Q.4. What all investigations you would like to recommend in this case to confirm your diagnosis?

Q.5. What is the mechanism of action of allopurinol in treatment of hyperuricemia?

Q.6. Enumerate various condition which may be associated with hyperuricemia.

Explanations

Ans.1. This patient most likely is having **acute attack of gouty arthritis.** Gout is a metabolic disease characterized by acute or chronic arthritis which is due to deposition of **monosodium urate (MSU) crystals** in joints, synovial fluids and soft tissues which is called **"Tophi".**

Classically described as a disease associated with hyperuricemia which may be due to metabolic defect in uric acid production or due to impaired renal excretion.

This disease has strong genetic predisposition and tends to affect elderly men and postmenopausal women more commonly.

"Podagra" is the name given when the joint involved is **big toe (foot)** and **"Gonagra"** is the name given when the joint involved is the **knee joint.**

Ans.2. Uric acid which is responsible for this disease is the end product of degradation of purine nucleotide in human. Steps of degradation of purine nucleotide are shown in Fig. 1.

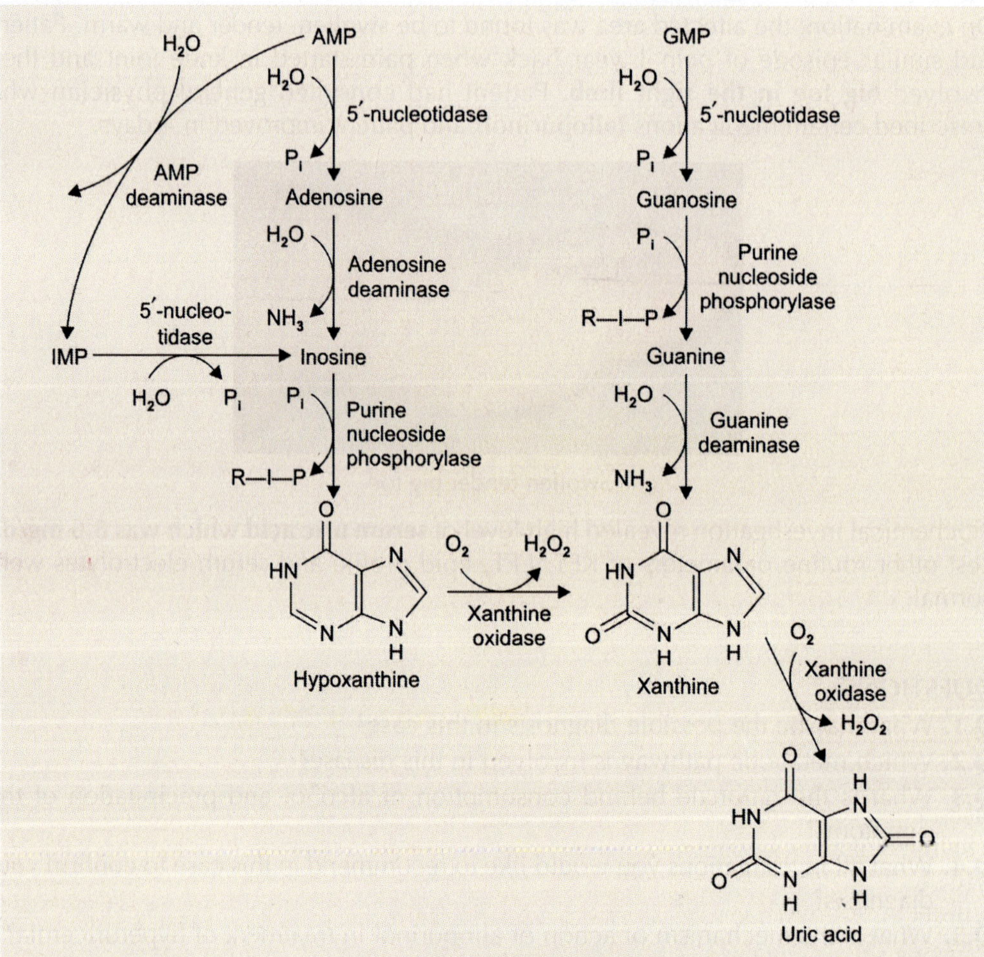

Fig. 1: Biochemical pathway of purine nucleotide catabolism to ultimately produce uric acid

Hyperuricemia may be due to either excess production of uric acid during purine nucleotide catabolism or it may be due to impaired excretion of uric acid due to lactic acidosis or due to impaired kidney function.

Ans.3. **Alcohol metabolism** in liver produces excess amount of NADH which converts pyruvate to lactate, hence **lactic acidosis** is seen. Lactic acidosis induced by alcohol metabolism impairs the uric acid excretion which is responsible for precipitation of the symptoms of gout.

Ans.4. To confirm the diagnosis, serum uric acid should be measured which often is elevated. Learner should always keep in mind that level of serum uric acid may be found within normal range or even low at the time of acute episode.

This is due to precipitation of excess uric acids in the joint in the form of uric acid crystals and also because of **inflammatory cytokines play uricosuric role, hence decreasing the level of uric acid in plasma.**

Hence, the elevated uric acid level in serum favors the diagnosis of gout but normal or low level of serum uric acid does not rule out the diagnosis of gout.

Gold standard of diagnosis is **visualization of MSU (monosodium urate) crystals** in the synovial fluid. MSU crystals are **negatively birefringent crystals.**

They show characteristic color when exposed to plane polarized light. When the axis of the MSU crystal is parallel to the polarizer, it appears **yellow.** When it is perpendicular, it appears **blue** (hence the name given is birefringent).

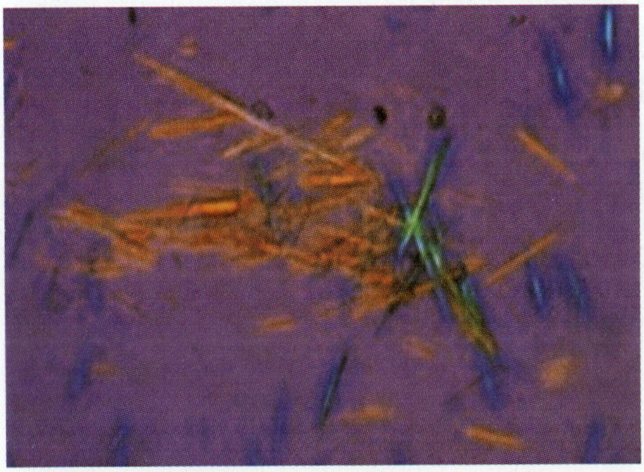

Fig. 2: Negatively birefringent crystals of monosodium urate (MSU)

Ans.5. The action of allopurinol on xanthine oxidase is a typical example of **"mechanism based inactivation (suicidal inhibition)"**. In this type of inhibition, the enzyme which is being inhibited is preparing its own poison, which means the enzyme activity is required to convert a less potent inhibitor to more potent inhibitor; which in turn inhibits the enzyme activity.

Classic example of suicidal inhibition is action of allopurinol which is being given in a case of overproduction hyperuricemia to reduce the production of uric acid. First of all allopurinol is converted to **alloxanthine by enzyme xanthine oxidase,** alloxanthine in turn inhibits the action of xanthine oxidase in a competitive manner. [Alloxanthine competes with the xanthine for binding the active site of xanthine oxidase].

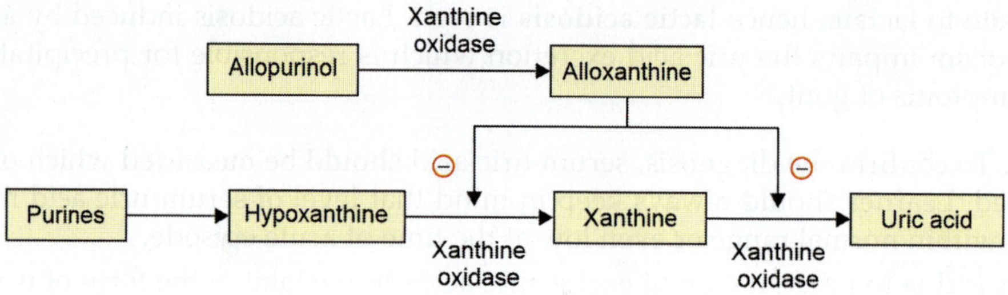

Fig. 3: Mechanism by which allopurinol inhibits xanthine oxidase

Ans.6. The conditions which are associated with hyperuricemia (overproduction of uric acid) are

a. von Gierke disease (refer to case 9)
b. Lesch-Nyhan syndrome (refer to case 92)
c. PRPP synthetase overactivity
d. Fructose intolerance (refer to case 3)
e. Galactosemia (refer to case 1)
f. Chemotherapy and radiotherapy.

Bibliography

1. Dhoble, A., Balakrishnan, V., Smith, R. Chronic tophaceous gout presenting as acute arthritis during an acute illness: a case report. *Cases Journal* 1, 238 (2008). https://doi.org/10.1186/1757-1626-1-238.
2. Soldo-Juresa D, Morović-Vergles J, Cunović S, Galesić K, Trutin-Ostović K. Prikaz bolesnika s gonagrom [Case report of a patient with knee gout]. Reumatizam. 2000;47(1):20-4. Croatian. PMID: 11552605.

CASE

92

Lesch-Nyhan Syndrome

A 4-year-old male child is brought to the hospital with complain of habit of biting of lips and fingers and involuntary movements of upper limb. On examination, baby had **multiple marks** of biting on fingers and lips.

Mother gives the history of **orangish discoloration of diaper** when the child was young. He is born on full term by normal vaginal delivery. He has younger sister who is healthy. **Serum uric acid was found to be 9 mg/dl (normal is 7 mg/dl).**

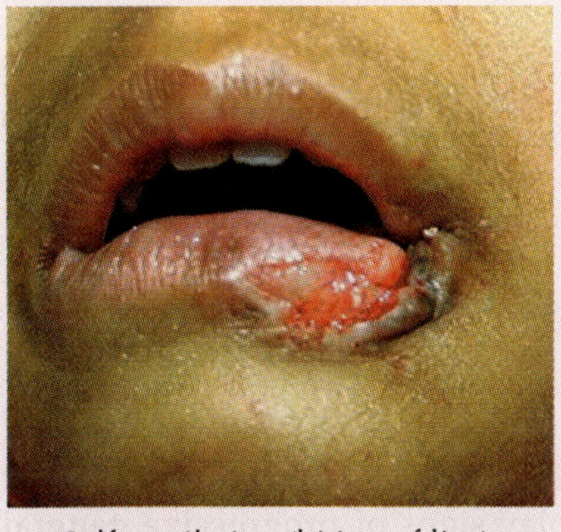

Self-mutilation (biting of lips)

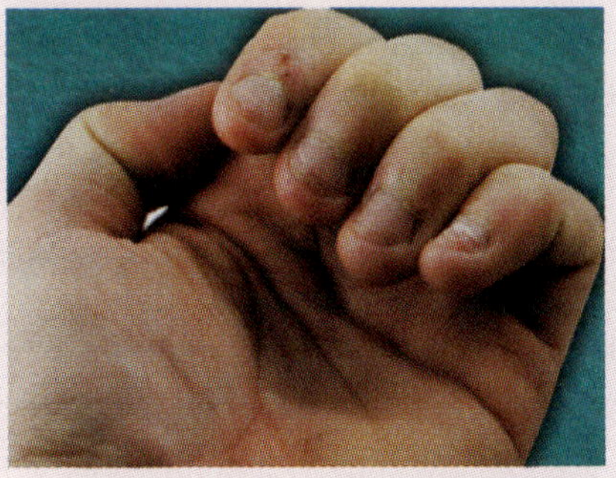

Self-mutilation (biting of nails)

QUESTIONS

Q.1. What may be the probable diagnosis in this case?

Q.2. What is the pathway affected in this disease?

Q.3. What is the reason of orangish sand found on diaper by mother?

Explanations

Ans.1. This baby may be suffering with **LNS (Lesch-Nyhan Syndrome).** It is an **"X-linked recessive"** disorder of purine metabolism which affects male child exclusively. In this disorder individual demonstrates spasticity, choreoathetosis, elevated urine, uric acid, self-injurious, and aggressive behavior.

Ans.2. This syndrome is associated with deficient activity of the enzyme **hypoxanthine-guanine phosphoribosyltransferase (HGPRT).** HGPRT is the enzyme which is used in salvage route of purine nucleotide biosynthesis, which converts hypoxanthine and guanine to inosine monophosphate (IMP) and guanosine monophosphate (GMP), respectively.

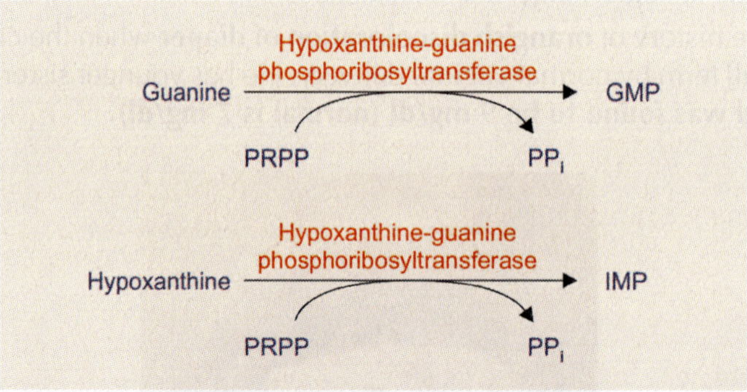

Fig. 1: Salvage pathway for purine nucleotide biosynthesis and role of HGPRT

In deficiency of this enzyme, excess hypoxanthine is diverted for degradation and uric acid formation.

At birth, most individuals with LNS show normal motor development, but by 6–8 months begin to display signs of **progressive generalized spastic paresis* and bilateral athetosis*.**

Ans.3. History of passing orange sand is characteristic history given by mother which denotes excretion of excess uric acid in the urine. These **orangish sands are urate crystals.**

Bibliography

1. Benz CM, Reeka-Bartschmid AM, Agostini FG. Case report: the Lesch-Nyhan syndrome. Eur J Paediatr Dent. 2004 June;5(2):110–4. PMID: 15202925.
2. Kale A, Shah K, Hallikerimath S. Lesch-Nyhan syndrome: a case report. J Indian Soc PedodPrev Dent. 2008 Jan;26 Suppl1:S11-3. PMID: 18974538.

Word Meaning

Spastic paresis: Spasm and muscle paralysis.

Bilateral athetosis: Involuntary movements of fingers, hands, toes and feet.